AGRITOURISM

AGRITOURISM

Michał Sznajder, Lucyna Przezbórska and
Frank Scrimgeour

www.cabi.org

CABI is a trading name of CAB International

CABI Head Office
Nosworthy Way
Wallingford
Oxfordshire OX10 8DE
UK

Tel: +44 (0)1491 832111
Fax: +44 (0)1491 833508
E-mail: cabi@cabi.org
Website: www.cabi.org

CABI North American Office
875 Massachusetts Avenue
7th Floor
Cambridge, MA 02139
USA

Tel: +1 617 395 4056
Fax: +1 617 354 6875
E-mail: cabi-nao@cabi.org

A catalogue record for this book is available from the British Library, London, UK.

Library of Congress Cataloging-in-Publication Data

Sznajder, Michal.
    Agritourism / Michal Sznajder, Lucyna Przezbórska, Frank Scrimgeour.
        p. cm.
    Includes bibliographical references and index.
    ISBN 978-1-84593-482-8 (alk. paper)
  1. Agritourism. I. Przezbórska, Lucyna. II. Scrimgeour, Frank. III. Title.

    S565.88.S96 2009
    338.4'779656–dc22

                                                                    2008025071

ISBN-13: 978 1 84593 482 8

Typeset by AMA DataSet Ltd, UK.
Printed and bound in the UK by the MPG Books Group.

The paper used for the text pages in this book is FSC certified. The FSC (Forest
Stewardship Council) is an international network to promote responsible management
of the world's forests.

Contents

About the Authors

Michał Sznajder has been a professor of agriculture and food economics at the Poznań University of Life Sciences, Poland, for 40 years. His numerous agricultural travels round the world have been the main inspiration for writing this book. His knowledge of agritourism has both practical and theoretical dimensions. He has visited numerous countries of the world and been invited by foreign universities and organizations to give lectures on different aspects of agricultural economics. He has published several books. His researches cover econometrics, the economy of the world dairy sector, neuromarketing and agritourism.

Lucyna Przezbórska, PhD, is a lecturer at the Poznań University of Life Sciences, Poland. Her main interests of study are connected with rural development, agritourism, rural tourism, agricultural economics and the economics of the world agriculture. She has led several research projects on agritourism in Poland and has also visited and studied rural tourism and agritourism enterprises in different countries of the world. She has numerous publications on rural tourism and agritourism.

Frank Scrimgeour is currently dean (and professor) at the Waikato Management School, University of Waikato in Hamilton, New Zealand. He has a long history of research on the economics of agriculture and the environment. He is particularly interested in the economics of pollution control, the efficient management of natural resources and the economic performance of the dairy and kiwi fruit sectors with ongoing work relating to agricultural and horticultural land values, international agribusiness strategy, analysis of nitrogen control strategies and the interactions between energy prices and the New Zealand economy. He has a variety of publications in major journals.

Preface

The history of recreation on farms, i.e. what is presently referred to as agritourism, is very long. It may well be as long as the history of towns and populations involved in non-agricultural business activity. Some authors indicate specific dates and forms of its initiation, e.g. Wolfe and Holland (2005) reported that in the USA farm-related recreation and tourism can be traced back to the late 1800s when city folk visited farming relatives in an attempt to escape from the city's summer heat. We remember from our young years wonderful holidays we used to spend in the country, visiting farms belonging to our relatives. That was true agritourism with all the attractions available then: sleeping in a haystack; simple but delicious food such as boiled new potatoes with butter and sour milk; endless hours spent playing with village children; wooden huts with thatched roofs; fields overgrown with countless crop species, peas, broad beans, beans, potatoes, oats, rye, clover; roaming red mountain cattle, turkeys, guineafowl, hens and ducks, or a beautiful stream to bathe in; going in a horse-drawn cart to and from the railway station; watching a foal running about. Who, back then in the 1950s, would have called it agritourism? Our parents would tell us about their own holidays spent in the country. That is the way it was then and still is all over the world, not only in affluent countries.

Modern times commercialized this type of recreation. Agritourism is now a dynamic business sector in many parts of the world. This is a significant change because until recently agritourism appeared as a backward activity lacking significant investment and dynamism. At present it has become the primary activity of numerous economic entities, not only agritourist farms, with a wide range of diverse forms. This book presents the depth of the new wave of agritourism apparent around the world. This provides entrepreneurs and policy analysts with a rich information set as they consider agritourist possibilities.

This book is dedicated to a wide range of readers: first of all to farmers engaged in this type of economic activity or considering it as their future career so that they can see how many different solutions are possible, how many people are interested in agriculture and the country, how satisfying this activity may be, both as a result of contacts with tourists and as a successful business operation. Readers could also be all those people who support the development of agritourism; those working for the state, regional and local administration and first of all extension services workers. This book is also addressed to the developing agritourism services, especially opening agritourism agencies and associations. We wrote it also bearing in mind all students of tourism, business studies, agriculture and forestry, social policy, anthropology or even medicine, as well as all those planning to work in rural areas. This book was also written for the sake of researchers interested in this problem. Finally, we would like to recommend it to prospective agritourists planning to spend their holidays in the country. They will surely find here plenty of interesting information and examples that will make their holidays in the country even more fascinating.

This book consists of four parts. The first section is an introduction to the economics of agritourism; it introduces the role of organization, management, logistics, safety, marketing, finance, economics and psychology and, within agritourism, discusses the role and importance of agritourism in the multifunctional development of farms and rural areas, showing the economic and social importance of agritourism. Furthermore, it highlights the problem of agritourist space. The second section focuses on the economics and organization of agritourism. This part presents a number of agritourism-specific concepts such as: the concept of an agritourist enterprise, economics and organization of agritourist farms, integration of agritourist entities, agritourism versus food processing, agritourist products and services, characteristics of tourists and the market for agritourist products and services. The third section is a specific case study dedicated to core agritourism, agritourist enterprises and related mobility issues. The cases come from 19 countries of the world, namely: Australia, Argentina, Belgium, New Zealand, Canada, France, Germany, Hungary, India, Ireland, Italy, The Netherlands, Poland, Russia, Switzerland, UK, Uruguay, USA and Zanzibar (Tanzania). In view of the fact that the authors come from two distant countries, Poland and New Zealand, it is understandable that examples from these two countries will predominate. This section concludes with a consideration of the agritourist potential of five countries of the world. The last, fourth part of the book opens for reconsideration and for discussion of several significant topics that appeared to be keystones for further agritourism development. At the end of each chapter, Case Studies are presented, which are to encourage readers to continue their investigations, suggest ways to adapt their newly gained knowledge to local conditions and facilitate the application of this knowledge in business operations.

Agritourism has an international dimension. Similar phenomena occur in various regions of the world. Comparative agritourism reveals how they adapt to various geographical and economic conditions. The whole book is enriched with photographs of farms, entities and agritourist enterprises. Many of the photos, especially in Parts I and II, are connected with comparative agritourism, which is extremely important for the international readership.

This book was written because no suitable book on the subject exists.[1] The book derives from the significant experience of all the authors in the analysis of agricultural and tourist issues in Poland, New Zealand and around the world, which has been facilitated by the University of Life Sciences in Poznań and Waikato University. All the authors contributed to each chapter, though Michał Sznajder wrote the major part of the book. Frank Scrimgeour gave this book its unique character to suit the international reading public.

This book would never have been written but for the invaluable graphic, editorial and research assistance of several people, especially Beata Moskalik and Jakub Majorek from the University of Life Sciences in Poznań.

We believe that this book will enrich the international literature on the subject and will also be an inspiration for many agritourist entrepreneurs searching for new sources of income, for policy analysts considering development opportunities and for city dwellers searching for interesting tourist experiences.

Michał Sznajder
Lucyna Przezbórska
Frank Scrimgeour
May 2008, Poznań, Poland, and Hamilton, New Zealand

---

[1] Many textbooks in agritourism have already been published. The following are available in English: Getz and Page (1997); Sharpley and Sharpley (1997); Butler *et al.* (1998); Hall, M.C. *et al.* (2000, 2003); Roberts and Hall (2001); *Agritourism in Europe* (2004); Eckert (2004); Hall, D. *et al.* (2004); Hall, D.R. *et al.* (2005). This book differs from those listed in the range of examples and concepts employed in exploring the practice of agritourism.

I

An Introduction to the Economics of Agritourism

1   The Concept of Agritourism

Definition

In the last 25 years of the twentieth century the term *agritourism* appeared in international literature. There exists a parallel word, agrotourism. The two terms have the same meaning; however, agrotourism sounds strange in the ears of the native speaker of English. Agritourism is more popular than agrotourism. An analysis of the number of entrances in Google shows that agritourism is used twice as frequently as agrotourism. Both terms consist of two parts: *agri-* or *agro-* and *tourism*. The prefix *agri-* derives from the Latin term *ager* (*agri* – genitive), which means 'field', while *agro-* comes from the Greek term *agros*, which means 'soil', and *agronomos*, which refers to a person managing a land estate, while *tourism* is a form of active recreation away from one's place of residence that is inspired by cognitive, recreational and sport needs – it comprises all forms of voluntary changes of one's place of staying.

The combination of the prefix *agri-* with the noun *tourism* resulted in the formation of a new word that means human tourist activity whose aim is to familiarize oneself with farming activity and recreation in an agricultural environment (Fig. 1.1). Agritourism is a style of vacation that is normally spent on farms. The term agritourism is understood differently by tourists and providers of agri-tourist services. For a tourist, agritourism means familiarizing oneself with agricultural production or recreation in the agricultural environment or it may include an opportunity to help with farming tasks during the visit. However, this definition does not fully render what the term agritourism means to people providing agritourist services. In fact, agritourism is a term introduced by representatives of the supply party representing the interests of farms providing agritourist services. This resulted in a considerable extension of the term to all activities related to providing services for tourists and holidaymakers. Therefore the entities providing agritourist services include in the term agritourism various forms of the accommodation industry – agri-accommodation; the food and beverage industry – agri-food

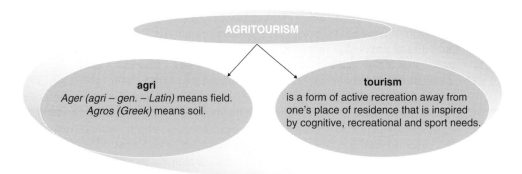

Fig. 1.1.   Etymology of the term agritourism.

and beverages; recreation – agri-recreation; relaxation – agri-relaxation; sport – agrisport; and even health care and rehabilitation – agri-therapy.

These terms are not new. Most of them are already used in practice. Moreover, agritourism includes certain elements of direct sales and any type of participation in or observation of agricultural production, i.e. agritourism proper. New portmanteau words have been coined, such as agritainment by combining the words 'agri' and 'entertainment' (Blevins, 2003). Most frequently for entities offering agritourist services and products agritourism is a mixed activity, which comprises all or some of the previously mentioned activities.

Farm management specialists point out that in general each farm by its nature has free resources, which are not used in the process of agricultural production. Such resources are: free rooms, own food (which the farmer's family is not able to eat, particularly during summer), free manpower resources, free space, and environmental resources like landscape, clean air, water reservoirs, etc. When used they give farmers an additional income. The resources may be used mainly in the tourism process. Therefore, they constitute natural touristic resources of the farms.

What underlies agritourism is the conviction that a farm is the basic entity providing tourist services. Having underutilized resources of labour, houses or cheap food, it can enhance its income by agritourist activity. However, practice shows that other business entities are also more and more interested in agritourist activity, while farms themselves, especially those located in areas predisposed to tourism, often narrow down their farming activity in favour of tourist activity. Agritourism is developing into a large part of the tourist industry and will probably soon be one of the largest sectors of tourism in some regions. There are instances of international investment penetrating the agritourist sector with quick recovery of the capital. Investments are situated in the most picturesque areas. It appears that farmers may be overtaken by other providers of the most attractive agritourist services and products, at the most attractive locations. They will only be providers of land and work dictated to by capital. Perhaps the least attractive branches of agritourism will be left to farmers. The *New Zealand Herald* of 26 November 2003 announced that an American multimillionaire, J. Robertson, had purchased part of the farm area comprising the picturesque cliff of Kauri in the north of New Zealand, not only investing in a hotel and golf course but also planning to create the largest private conservation area in New Zealand. However, some of the investor's plans, especially the idea to bore into the cliff space for a restaurant with a view of the sea, have set off protests by the local community.

The range of the agritourism term, just as many other terms, significantly changes over time. The traditional, rather passive apprehension of agritourism concerns almost exclusively just vacationers staying on farms and what farmers can offer to the vacationers within the frame of the farms' natural touristic resources. Traditional agritourism gave small, additional income for the farmers. Modern agritourism is understood as the more active one. The farmer, initially, invests in agritourism, broadens the offer of the agritouristic products and the income obtained from agritourism consists of a large, or even an exclusive, part of the farm income.

Agritourism for the new generation is appearing on the horizon. Tourists want to experience agriculture and rurality in a new more intensive manner. Actually the typical farmer will not deliver services for tourists, but probably the former farmer transferred into high class specialist agriculture and tourism. He will not be the owner of the farm anymore but he will become the owner of an agritouristic venture. Agritourism of the future becomes a modern and lucrative business based on the latest information technology.

Agritourism vs Rural Tourism

Agritourism is not the only area within tourism. Classifications yield many other distinguished fields, such as ecotourism, garden tourism, guest ranch, safari, village tourism, wine tourism, dairy tourism and, of course, rural tourism, which are related to agriculture, forestry, food processing and rural areas. Rural tourism is not a term identical to agritourism. However, the two terms are very strongly interlinked. Besides being connected to plant and animal production and processing, rural tourism usually also comprises those types of human activity that are related to living in the country, its culture, religion and everything that is understood by the term ethnography (or ethnology), i.e. a scientific discipline whose subject of research is man as the creator of culture. Ethnography recognizes an autonomous sphere of human existence that decides about the individuality and distinct character of ethnic groups. It is unnatural to concentrate only on agritourism, excluding rural tourism. Farms are, in fact, a crucial part of country areas.

The range of the term agritourism varies depending on the geographical region of the world. The relations between the terms agritourism and rural tourism also change regionally. This variability is the result of the role that agriculture and rural areas play in a particular country or region. In many American states, where a rural community in the European sense of the word is almost non-existent, the terms agritourism and rural tourism are almost equivalent. Dinell (2003) characterizes the term agritourism in the state of Kansas as follows: 'agritourism comprises tourist trips to farms and ranches, the trips which are made up of packages of events in country areas and trips to food processing plants'. In Europe, where rural areas fulfil a number of non-agricultural functions, distinguishing between agritourism and rural tourism is relevant. In the areas where farming production has disintegrated but where a strong rural community exists, the differentiation between rural tourism and agritourism seems to be justified. In view of the fact that the importance of traditional rural communities is being narrowed down whereas the importance of cities is growing, it is likely that in the future the term agritourism will be used more frequently than the term rural tourism. Specific situation concerns the states of the former Soviet Union, where in the collectivization process small, private farms and traditional rural communities were almost completely destroyed. Therefore agritourism and rural tourism are completely new concepts and they will have to be formed from the very beginning. Figure 1.2 shows a pyramid giving a hierarchical positioning of rural tourism and agritourism in relation to each other and in relation to mass tourism and alternative tourism. The higher in the pyramid a particular type of tourism is positioned, the narrower the range of its focus.

The lack of precision in defining the terms agritourism and rural tourism makes it difficult to clearly specify the range of this book. In principle, the book concentrates on the economics of a farm providing agritourist services. Given the close practical connections, it will refer to rural tourism and the agri-hotel industry, agrifood and beverage, agri-recreation, agri-relaxation, agrisport and agri-therapy. In its broader sense agritourism also comprises some services provided by food-processing plants and companies trading in farming products. This aspect of agritourism is important for promotion of the companies' products in countries with a well-developed food industry.

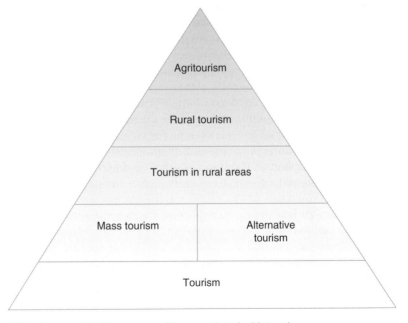

Fig. 1.2.   A pyramid of the range of terms related with tourism.

In view of the difficulty of clearly defining the terms associated with various types of tourism, a question arises as to whether distinguishing agritourism from tourism is not artificial. After all, both tourism and agritourism aim at satisfying the tourist's demand for sightseeing, learning, relaxation, recreation or sports. Is not tourism in many countries carried out mainly in rural areas? Are there any features that differentiate tourism from agritourism? If there are some distinguishing features, then differentiating agritourism is justified. The answer to this question is at least partly positive. However, more and more frequently types of branch tourism are distinguished from tourism, e.g. tourism related to wine production – wine tourism, food tourism, ecotourism, etc. It is a fact that tourists visit farms and rural areas. Therefore, there are no reasons why agritourism and rural tourism should not be distinguished. Below are three main features that differentiate agritourism from conventional tourism.

Uniqueness of Agritourism

The first feature is the possibility to satisfy human need with practical participation in the process of food production, in the life of a rural family and in a rural community. The tourist has a chance not only to participate in plant and animal production and food processing but also to take part in the life of a farm family. Agritourism understood in this way is a difficult but very ambitious form of tourism. This form of tourism is not interesting to all tourists but only to those who want to link relaxation with acquiring new practical skills or experiences.

The second characteristic quality of agritourism in relation to conventional tourism is the possibility to satisfy the human cognitive need within farming production or ethnography. Agritourism gives a chance to learn about the lives of rural people, their culture and customs.

The third feature of agritourism is the possibility to satisfy emotional needs, which is the willingness to have direct contact with domestic animals, plant and animal products and the products of processing, and the need to experience the idyllic countryside associated with the atmosphere of rusticity, silence, sounds or even smells of the country and farm. Agritourism void of the cognitive element, eliminating human emotional needs and limited only to relaxation does not differ significantly from conventional tourism.

The question arises as to whether agritourism should be recognized as a new scientific discipline. In order to recognize a certain area of science as a new discipline, it should comprise a specific range of knowledge, the subject of the research should be clearly defined and the discipline should use specific methods of its own. As has already been shown, the range of interest of agritourism is not clearly defined; however, it is possible this will happen in the future. So far, there have been no methods that could be defined as specific to agritourist research. One may suppose that in the future research methods specific to agritourism will begin to appear, the signal of which is Drzewiecki's method of evaluation of agritourist areas (Drzewiecki, 1992). Clear definitions and developing specific research methods would raise agritourism to the rank of a scientific discipline. Currently it is applying methods taken from statistics, economics, agronomy, animal science, zoology, geography, ethnography and many other scientific disciplines. At the moment, distinguishing agritourism as a scientific discipline is premature. At present, agritourism should be treated as a human business activity that is described and examined by such scientific disciplines as economics, geography, biology, agriculture, law, ecology, ethnology and others (Fig. 1.3). Agritourism is a real and constantly developing human activity. Reports of this activity come from many parts of the world, including nearly all American states, Canada, all western, central and southern Europe, Australia and New Zealand and even Asia and Africa. The development of agritourism is induced by both the demand and the supply. A key factor is the growing interest of city inhabitants in where food comes from and how it is produced.

Agritourism as a subject is taught at various universities. A question arises as to which departments and universities are competent enough to teach classes and conduct research in agritourism. The question is crucial because agritourism is developed in university departments of different faculties. So far it has been dealt with by academics from the departments of economics, geography, agriculture, animal breeding, tourism and landscape design from different universities. Finally, it is appropriate to consider the economic importance of agritourism. The living standard of rural people in many countries of the world is very low. Research conducted by economists in the last decades has shown that a considerable number of people living in rurally populous countries are not able to earn a living from agricultural production, that is, from plant and animal production. Too little income can be generated from this production to satisfy the people's needs. It is necessary to gain income from non-farming sources. In many countries of the world farmers with no chance to satisfy their material needs at the place of residence move to a city, in which as often as not they also fail to find employment. The consequences of such migrations are deplorable. Slums are formed, which are conducive to the emergence of social pathologies of different types. Economic and social policies, especially in the countries of the European Union, aim at keeping a considerable number of village people in rural areas. The necessity to satisfy the financial needs of people involved in plant and

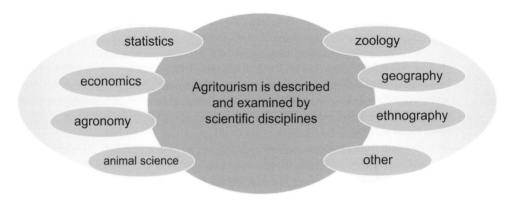

Fig. 1.3. Multidisciplinary character of knowledge in agritourism.

animal production caused many farms to start non-farming activities, including agritourism. Agritourism is a crucial element of the social policy of countries of the European Union and this is known as the multifunctional development of farms and multifunctional development of rural areas. Agritourism is an example of non-agricultural development of farms and rural areas.

Thus agritourism comprises a wide range of knowledge, which can be presented in terms of economics, organization and management.

Functions of Agritourism

Among numerous meanings of the term function, the basic definition taken from mathematics is used most commonly. In this way functions are treated as a set of clear-cut relationships found between two or more sets of elements. Thus the functions of agritourism result from diverse relationships and relations, formed at different stages in the development of agritourist activity. Thus it may be assumed that functions of agritourism result directly from the consequences (effects) of its development and pertain to different aspects of the functioning of socio-economic systems. The large number of functions ascribed to tourism, including also agritourism, results from the fact that it affects many diverse aspects of human life.

Agritourism serves several functions of highly varied importance, both positive (eufunctions) and negative (dysfunctions). Thus we may distinguish functions connected with revenue, employment, utilization of available housing facilities, activization of rural areas, nature conservation, as well as those related to rest, recreation and education of both urban and rural populations. Functions of agritourism may also be viewed in terms of three categories – as expected functions (expected effects of the development of agritourism), as postulated functions (desirable effects of the development of agritourism) and actual effects of its development (Gaworecki, 2006).

Commonly three agritourism functions are found in the literature (Fig. 1.4):

- Socio-psychological;
- Economic;
- Spatial and environmental (Iakovidou *et al.*, 2000).

The first, i.e. socio-psychological, functions are connected with increased respect for the rural community, the intermingling of rural and urban cultures and an opportunity to enjoy contacts with the traditional lifestyle of the rural community. Socio-psychological functions of agritourism may include:

1. Gaining new skills, experiences and professions, learning foreign languages, gaining entrepreneurial skills, activization of the rural community, formation of new capacities in tourist services, broadening one's knowledge or learning more about one's local area, its history and attractions, encouragement of social initiatives or new opportunities for rural women.

2. Meeting new people, a possibility to make new contacts and social ties, exchange of experiences or attitudes, on the part of both farmers and their guests, increased tolerance in relation to different attitudes, behaviour or opinions, broadening of knowledge on the world and other people on the part of farm owners, encouragement to develop hobbies and interests.

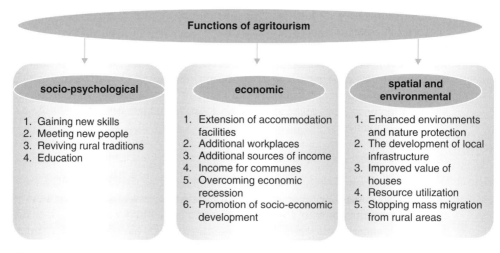

Fig. 1.4.   Functions of agritourism (based on Iakovidou *et al.,* 2000).

3. A possibility to revive rural traditions, promoting respect and revival of folk traditions and culture, the development of culture in rural areas, fuller utilization and revival of certain objects in villages (community centres, sports facilities, etc.).

4. Educational functions of agritourism are connected with learning about the real world (nature, cultural heritage), which modifies specific attitudes in relation to different aspects of reality (the host and guest, a group of tourists, family); agritourism is also a medium to express one's feelings (learning about and respect for farmers and farm produce); agritourism offers an opportunity for tourists to be creative (participation in farm work, learning a folk craft, etc.), contributes to good health (climatic conditions, food, exercise).

Economic functions concern the stimulation of development of agricultural, horticultural or animal-breeding farms, generation of additional sources of income both for rural households (frequently at limited investment outlays) and for local or regional governments and communes. The group of economic functions includes:

1. Extension of accommodation facilities, maintenance of existing production, extension of assortment and improved quality of offered services facilitate direct sales of certain farm produce, contribute to the formation and development of additional markets for foodstuffs and different types of local services, such as crafts, handicraft products, artistic metalwork, etc.

2. Creation of employment and reduction of unemployment rates, including latent unemployment, which results in inhabitants of villages being needed, socially accepted, encourages them to develop qualifications (the psychological aspect) and facilitates the utilization of the human resources potential.

3. Obtaining additional sources of income for farmers (increased revenue for farmers, and thus income, may be allocated to investment outlays, e.g. construction or renovation works), which results in reduced dependence on farming, diversification of the local economy, which in this way becomes less vulnerable to market fluctuations.

4. Obtaining additional income for business, communes, local governments of a given town, associations of communes or the region.

5. Overcoming economic recession (thanks to its interdisciplinary character, tourism activates different professional and social groups), additionally tourism is a revival factor in rural areas and the revitalization of the rural community by offering possibilities of social and economic advancement.

6. Promotion of the socio-economic development of underprivileged areas, diversification of economic activity in rural areas, creating conditions and opportunities for the development of other types of activity in rural areas.

Since agritourism in the process of development uses elements of the natural environment, transforming them, spatial and environmental functions include the consequences of the development of agritourism for the natural and anthropogenic environments. Environmental functions include:

1. Enhanced care for the environment, nature protection, creating a more friendly environment for guests and visitors.

2. The development of local infrastructure (water supply, sewage systems, sewage treatment plants, roads, public transport, recreation facilities), which makes life in the country easier and improves the standard of living for rural populations.

3. Improved aesthetic value of houses and areas in their vicinity, care for the aesthetic value of villages, houses, streets and other public spaces – aesthetic enhancement of villages.

4. The utilization of old, frequently derelict buildings (rarely used rooms, attics, whole uninhabited buildings, parts of households, farm buildings, windmills, restaurants, shops, castles, palaces, manor houses, etc.), which may contribute to the preservation of the rural cultural heritage.

5. Countering mass migration from rural areas (mainly of young and educated people) and the depopulation of rural areas (such trends could be observed in some regions of many European countries, e.g. Great Britain, Germany, Hungary, or in certain periods in Poland).

Some of the above-mentioned functions of tourism in rural areas overlap, supplement or result from one another. It is also difficult to define which of them are more and which less important. However, several studies have shown that the primary function of agritourism stressed by farmers and rural accommodation suppliers is related to offered income.

Aspects of Agritourism

Running agritourist activity requires knowledge in many areas. Only such knowledge may be comprehensively transformed into a product or service that may be offered to tourists. Each area of this knowledge is defined here as an aspect of agritourism. Aspects of agritourism include organization, management, marketing and economics.

Organization of agritourism

The scope of the term 'organization of agritourism' covers many issues. It pertains first of all to business and finance law, as well as safety regulations, creating a framework for agritourism. There are also specific regulations concerning agritourism itself. Next the organization pertains to structures functioning within it, i.e. farms, agritourism associations and institutions, as well as supporting institutions. Organization is related to forms of ownership and obligations. Organization of structures also includes vertical, horizontal and territorial franchising integration and joint action of these entities. An important element in organization is the position of a given economic entity in the agritourist space.

Organization concerns the internal structure of the agritourist entity, i.e. its internal structure and links between elements of this structure. An important element here is related to relationships between products and services offered by the farm. Thus competition is generated between products in relation to the farm resources, creating independent, complementary, linked and competitive products. All these organizational elements will be successively discussed in this book.

Management in agritourism

Management in agritourism concerns a wide range of issues that are relevant to the success of the enterprise. It includes the management of logistics, management of products and services, hospitality, quality and safety, as well as human resources. Logistics provides the influx of tourists. Management of products and services results in a situation when the farm is becoming increasingly attractive and thus capable of meeting the expectations of tourists. Hospitality with special emphasis on psychology facilitates an individual approach to tourists, ensuring their satisfaction with the agritourist offer, and includes the manner of receiving guests; conversation, preparing and serving meals, the artistic programme and even interior decoration and surroundings all belong to the realm of management. Among the most important challenges are safety and quality management and control. In many countries there are organizations that give agritourist farms quality signs depending on their standard.

Management is not the primary focus of this book. Nevertheless, the contents of the book include numerous references to this field, especially when discussing agritourist products, describing farms or analysing mobility.

Marketing in agritourism

Even a well-organized agritourist farm may not yield satisfactory financial results without marketing. In this respect marketing plays a crucial role. Advertising and promotion are indispensable. Advertising forms used today are especially promising. Online booking services are a definite necessity.

Marketing is also related to the pricing policy. Prices not only determine revenue, but also inform the target tourists.

The essence of marketing is a well-developed product. Agritourism offers a huge and unlimited range of products. Numerous examples are presented in this book.

Moreover, marketing analyses potential consumers, dividing them into segments in terms of age, sex, place of residence, income level, preferences and especially lifestyle. As there are practically no such studies on the segmentation of agritourists, it is going to be the subject of a separate chapter in this book.

Location of an agritourist farm also plays a role, especially in terms of the agritourist space in which it is found. Although this book is not a regular lecture on marketing, it contains information in this field.

Economics of agritourism

Agritourist activity has an economic aspect; hence we can speak of the economics of agritourism. As with agritourism itself, the focus of the economic analysis of agritourism is not precisely defined. This book attempts to define a framework for the economics of agritourism. The economic challenges in agritourism include consideration of the regional and national economy, the economic decisions of firms, consumers and policymakers, economic policy, the economics of production and marketing and town and country planning. The economic analysis of agritourism describes and analyses entities functioning in the sector, which are not only farms but also enterprises and associations.

Agritourism is an object of interest for social and welfare policies of countries. The idea is that rural people should gain increasing income from non-agricultural activity. The market of agritourist services is developing and the analysis of it includes the demand, supply and structures facilitating market processes. The economics of agritourism also includes segmentation of consumers of agritourist services and market analysis. It includes the problems of financing agritourist enterprises, investing in the activity and rural people's income.

Case Studies[1]

1. Define the joint and individual scope of the following terms: agriculture, rural areas and tourism.
2. List similarities and differences in the terms 'rural tourism' and 'agritourism'.

---

[1] At the end of each chapter, tasks referred to as 'Case Studies' are given, which are to serve three objectives: study, adaptation and application. The first objective – study – consists in the consolidation and broadening of knowledge on agritourism and an indication of directions for further self-study. The second objective – adaptation – consists in the adaptation of this knowledge to local conditions experienced by the reader. It results from the vast diversification of the economic, agricultural and natural environment worldwide. This adaptation to local conditions will be the original developmental

3. Which functions of agritourism in your opinion are of greatest importance for agriculture and which for the development of tourism?
4. Figure 1.4 presented groups of individual functions of agritourism. Discuss these functions.
5. On the basis of a selected region or farm, discuss the practical implications of the spatial management function. Decide whether and how the function of spatial management may lead to development or stagnation in agritourism.

input by the reader. The third objective – application – consists in the specification of knowledge contained in this book so that it may be applied in practical operations at the macro (i.e. regional) level and the micro (a specific farm) level. At the regional level this knowledge may be used in planning the development and promotion of agritourism, while at the micro level it will facilitate the preparation of a business plan or restructuring of a given agritourist farm. In order to effectively meet the third objective the reader should at the very beginning define the area of the region and specific farms which adaptation and application tasks will refer to.

2 The Economic and Social Importance of Agritourism

The Context of Agritourism Development – Low Farm Incomes

The economic and social importance of agritourism can be considered in many aspects, e.g. the macroeconomic aspect – the importance of agritourism for the economy of the whole country, the regional economic aspect – the importance for the local community and the microeconomic aspect – the importance for agritourism entities providing products and services and for the receivers of those products and services, i.e. their consumers (agritourists). The primary phenomena that affect the development of agritourism are:

1. Low farm income;
2. Urbanization;
3. Redistribution of urban people's income to agritourist farms and in consequence to all rural people;
4. The state of rural infrastructure and level of urbanization;
5. Policy of local self-government.

The history of the development of agriculture and rural areas shows that the people inhabiting rural areas have usually been traditional, religious, hard-working and rather poor. *The Angelus*, painted by Jean-François Millet in the years 1857–1859, whose reproduction is shown in Fig. 2.1, reflects the traits of the people. Usually the people who were occupied in non-farming activities were richer and less hard-working.

History provides us with numerous examples of peasant insurrections resulting from their famine, misery and poverty: for example, the Yellow Turban Rebellion of AD 184 in China and during the 14th and 15th centuries about 50 peasant insurrections broke out in the German Empire. In Russia there were several peasant rebellions, e.g. Stiepan's rebellion (1667–1671) and the peasant war led by Jemeljan Pugatshov (1773–1775). In Galicia (Austria-Hungary), a peasant insurrection led by Jakub Szela broke out (1846–1847). There are

Fig. 2.1. The hard-working, poor and pious rural people are perfectly depicted in the painting by Jean-François Millet *The Angelus* (*L'Angélus*, 1857–1859 (RF 1887)) (musée d'Orsay, Paris, reproduced with permission).

numerous examples of peasant rebellions from different countries. The background of all the insurrections was hunger, misery and poverty. Peasants were severely repressed after unsuccessful insurrections. Most of them were violently suppressed and the insurrectionists were cruelly punished.

In history the peasants' material predicament and the socio-political tension resulting from it was defined as the peasant problem. In the 19th and 20th centuries the peasant problem was to be radically resolved. In the Soviet Union it was 'resolved' by force – by extermination of peasants instead of giving them a chance for development beyond agriculture.

The Income Situation of Farmers in the EU Countries

The problem of rural people's income, especially farmers' income, is almost universal. It applies to many countries of the world, including the countries of the European Union. Hence one of the main goals of the Union, which was written as early as the Treaty of Rome (1957), is 'to ensure . . . an appropriate living standard to farmers, particularly by means of increasing the income of people employed in agriculture'. The Community, which has a range of tools of fiscal and agricultural policy, takes various actions aimed at sustaining the farming income and in many member countries the income of farms is even higher or equal to the average income of all households. However, income from agricultural business has changed steadily since the end of the 1950s and now it is shared among many parties, and farm household income from off-farm sources, including off-farm work, investment and other sources, has increased dramatically.

Eurostat, besides measuring income from agricultural production, has set up a methodology for measuring the income of agricultural households (Income from Agriculture Household Sector (IAHS)). The main aggregate income concept used in the IAHS project is net disposable income, adapted from national accounts methodology. The net disposable income includes not only income from farming and other activities (from independent activity – self-employment – and from dependent activity – employment), but also from pensions and other forms of transfer (property, social and other transfers). It is after the deduction of items such as current taxes, social contributions and other payments. Eurostat uses two kinds of household definitions (Rural Households' Livelihood and Well-Being, 2007):

- An agricultural household ('narrow' definition) is one where the main income of the household reference person (typically the head of a household) is from independent activity in agriculture, i.e. farming; a range of other socio-professional groups can be established on the same basis for the purpose of comparison;
- An agricultural household (supplementary or 'broad' definition) is one where any member of it has some income from independent activity in agriculture.

According to the Eurostat calculations and findings the number of agricultural households where the main income of the head of a household comes from farming is substantially smaller than the number of households where there is some income from farming. Agricultural households (in the 'narrow definition') in all EU countries are recipients of substantial amounts of income from outside agriculture.

In EU countries typically about a half to two-thirds of the total income comes from farming, but there are significant differences between member states and some differences between years. The total income of agricultural households is more stable than their income from farming alone. Non-agricultural income (taken together) is less variable from year to year than is farming income. Disposable income seems to be less stable than total income, but the relationship between the two depends on a variety of factors, including the way that taxation is levied. Agricultural households have average disposable incomes per household that are typically similar to, or higher than, the all-household average, although the relative position is eroded or reversed when income per household member or per consumer unit is examined (Fig. 2.2; Rural Households' Livelihood and Well-Being, 2007).

The main exception to the finding that average disposable incomes per household are similar to or higher than the all-household average was Portugal, where it was less than half. Lower levels were also in Greece (86%) and Italy (90%), but also in Spain and Germany. The relative position is eroded when income per

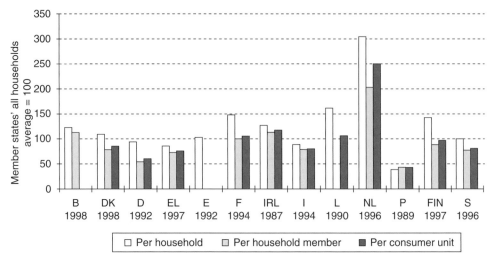

B, Belgium; DK, Denmark; D, Germany; EL, Greece; E, Spain; F, France; IRL, Ireland; I, Italy; L, Luxembourg; NL, The Netherlands; P, Portugal; FIN, Finland; S, Sweden.

Fig. 2.2. Average disposable income of agricultural households relative to the all-household average (selected member states). Note: For Luxembourg, in the absence of a comparison being generated within the IAHS statistics, interim figures taken from a survey of living standards have been substituted. (Redrawn from Income of the Agriculture Household Sector, 2001 report. Eurostat; Rural Households' Livelihood and Well-Being, 2007, p. 388.)

household member or per consumer unit is examined. Nevertheless, on all three mentioned measures, agricultural households had incomes at or above the national averages in France, Ireland, Luxembourg and The Netherlands. However, agricultural households on average usually had incomes lower than households headed by other self-employed reference persons in the same member state (Rural Households' Livelihood and Well-Being, 2007). Simultaneously the phenomenon of dual professions in farmer families and combining income from two or more sources is increasing. In the countries where multiple professions are a common phenomenon, e.g. in Germany, it is most frequently attributed to insufficient income from the agricultural activity and the surplus of work resources in the farm.

On the basis of research conducted in the farms of 'the old EU members' in 1987, it was found that less than a half of farmers' households gained more than 50% of their income from agricultural production (Bryden, 2000). The EU documents show that in the 1990s about 60% of the labour force working in agriculture gained extra income also from other, non-agricultural activities (Opinion of the Committee, 1997). At the same time, in farmer households on average one-third to two-thirds of the total income came from non-agricultural activity, and farming brought less than half of the total income of farmer families in Italy, Germany, Greece and Denmark. In Holland, Spain, Ireland, Belgium and Luxembourg, the share amounted to about 70%. More and more frequently, especially in Germany, Italy, Finland and Greece, farmer households gain a half or more than a half of their income from non-agricultural sources (Mahé and Ortalo-Magné, 1999). External income begins to dominate the structure of the income of farmers'

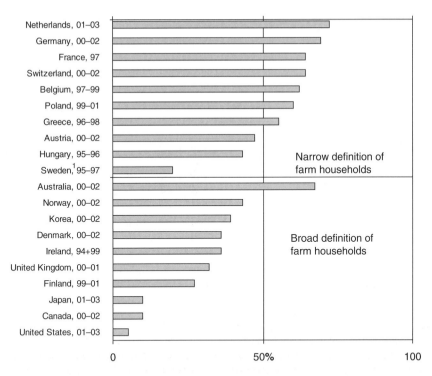

Fig. 2.3.   Percentage share of farm income in the total income of farm households in selected OECD countries (average of the three most recent years available). Note: Data are not comparable across countries.[1]Income from independent activities. (Sources: Update from OECD (2003); *Farm Household Income: Issues and Policy Responses*, Paris; Farm Household Income, 2004.)

households in those countries. In such countries as Ireland or Germany one in four people working in agriculture has a non-agricultural source of income, which is usually the main source of maintenance for them (Barthelemy, 2000). In Ireland 43% of farmers' households had at least one of the family members employed outside the farm gaining on average about 30% of the total income of these families (Keeney and Matthews, 2000). The average share of farm income in the total income of farm households in selected OECD countries is shown in Fig. 2.3.

Tendencies concerning the choice of individual sources change in time, and in different countries they have different strength. For example, in Germany in the 1970s farmers usually found additional employment in industry or building. In the 1990s there was a shift towards the services provided in one's own household, which were usually related to the processing of agricultural products, storage and sales of own products, providing tourist services or crafts. Simultaneously in some countries the importance of non-profit income is increasing. In Italy income from social services has the biggest share in the structure of the total income of an average household, exceeding the level of income from agriculture.

The transformations in employment and sources of income cause changes in the share of purely agricultural farms in the total number of farms. According to

Table 2.1. Share of part-time and paid workers' employment in agriculture in chosen countries of the EU, 2006. (Employment in agriculture and in the other sectors: structures compared, 2007).

Countries	Share of part-time employment in agriculture (%)	Share of paid workers in agriculture (%)
Austria	18.2	21.7
Denmark	21.0	57.1
Finland	19.1	32.2
Germany	22.8	62.2
Greece	11.3	8.8
The Netherlands	38.6	54.3
Ireland	n/a	22.9
Italy	10.6	52.6
Poland	24.0	12.6
Portugal	47.7	17.1
Spain	11.3	8.8
EU (27)	21.2	33.7

Eurostat, the agricultural sector provided a part-time job for 21.2% of the 12.6 million people working in the sector in 2006. Part-time employment is especially common in such countries as Portugal and The Netherlands (Table 2.1). However, the employment situation has changed over the years. Part-time agriculture in these countries is above all attributed to the structure of farms and structure of production, which contribute to an increased demand for workers only in particular seasons of the year. The proportion was more than 50% in Denmark, Ireland, The Netherlands, Austria and Finland (Rural Developments, 1997).

Although such phenomena as gaining additional income outside the farm and part-time agriculture are largely related to the structure and size of farms, types of agricultural activity and work efficiency, they are undoubtedly signs of the time and transformations occurring in agriculture. For many farmers' households agricultural income is becoming a component of the total income and this trend is most likely to be continued in future. More and more frequently, in households with multiple sources of income, agricultural activity is becoming an additional activity and sometimes it is even treated as a hobby.

According to the Eurostat data (Rural Development in the European Union, 2007) the diversification of the rural economy to sectors other than agriculture is progressing. In 2005, 36% of European farmers had a gainful activity other than agriculture; however, the share was even higher than 50% in such countries as Slovenia, Sweden, Cyprus, Malta, Denmark and Germany (Fig. 2.4). The same data show that 86% of employment and 95% of value added in predominantly rural areas of the EU (27), where over 50% of the population live in rural communities, came from the non-agricultural sectors and one of the main opportunities in terms of potential growth for rural areas comes from tourism, which plays a major role among non-agricultural activities in rural areas. It is estimated that

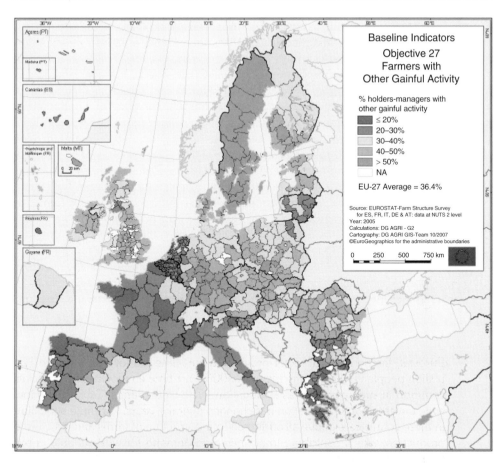

Fig. 2.4. Farmers with other gainful activity in the European Union countries (2005) (source: Rural Development in the European Union, 2007).

nearly 75% of bed places in the EU (27) are in rural areas (Rural Development in the European Union, 2007). The report on rural development underlines that there is a huge potential in the rural population and that this is the key factor for the development of rural areas.

Similarly to many other countries, Polish agriculture as a source of income has been subject to decline for a long time. In the last decade of the 20th century there was a notable decrease both in the agricultural disposable personal income (or net income) and in the income of farm households, especially when calculated per person working in agriculture. The system transformation of the 1990s enhanced the problems that have always nagged the country and agriculture, especially hidden unemployment and low work efficiency. Thus, it contributed to even more rapid deterioration of the economic situation of agriculture, including the income situation.

The General Census of Agriculture of 2002 showed that 363.4 thousand farms, i.e. 12.4% of all entities in the country, including 360.7 thousand individual farms, ran a non-agricultural business on their own account. In comparison with the previous census of agriculture of 1996, the number of farms running a non-agricultural business had risen by 86.9 thousand, i.e. by 31.4% (Fig. 2.5). In Poland, according to the Central Statistical Office, a farm is arable land, including forest land, buildings or their parts, appliances and livestock if they constitute or can constitute an organized economic entity, with the rights and duties related to keeping a farm. An individual farm is an agricultural lot of the area of 0.1–1.0 ha of agricultural land including the owners of farm animals, and an individual farm whose area of arable land exceeds 1 ha that is the property of or is used by the same person or group of people (Report on the Results of the National Agricultural Census 1996, 1998; Report on the Results of the National Agricultural Census 2002, 2003). There were nearly 257.4 thousand individual farms, i.e. 8.8% of the total number of farms, which ran an additional non-agricultural business along with the agricultural activity (in 1996 there were 167.1 thousand such farms, i.e. 8.1%).

The people who decided to run their own business were usually owners of big farms, whereas owners of small and poor farms are much less active in the trend and they would rather become hired labourers to earn money. The farm

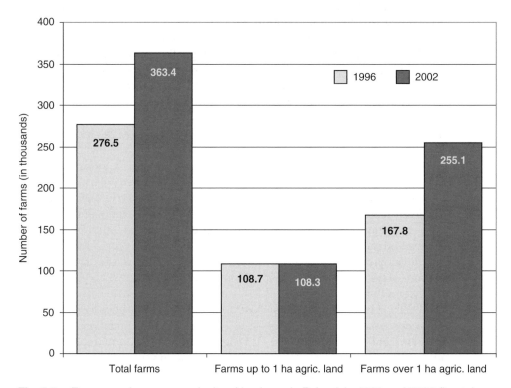

Fig. 2.5. Farms running a non-agricultural business in Poland, in 1996 and 2002 (based on Report on the Results of the National Agricultural Census 1996, 1998; Report on the Results of the National Agricultural Census 2002, 2003).

users who declared that they were running a specific type of non-agricultural business usually indicated an activity related to trade, agricultural and food processing, building and transport. In addition, the farms with businesses related to forestry (obtaining wood, hunting, forest growing, collecting, etc.), fishing and fisheries, ran restaurants, bars, canteens, administered real estates, rented machines and equipment without operating them, dealt with building (erecting buildings and structures, making building installations, doing finishing work, hiring out building equipment including operation), manufacture (i.e. production of food and drinks, manufacture of fabrics and clothes, leather tanning and dressing, manufacture of wood and wooden products, straw and wicker, paper products, rubber products and plastic products, waste disposal and manufacture of ready-made metal products and other products from other raw materials), hotel business and rented out rooms (running hotels, campsites, restaurants, bars, canteens and providing services for tourists). The number of non-agricultural business entities in the country is rising. These are mainly trade and service places and food-processing plants. Most of them are small plants, which require neither very large capital to start nor high qualifications. Farmers invest the means they have at their disposal and members of their families are the labour force. With appropriate orientation in the local market, the activity that businessmen-farmers decide to take on almost always brings extra income. In the case of the rural job market, such enterprises begin as family businesses and, as they develop, they provide new employment opportunities.

The most frequently listed reasons for the fall in farmer income in the 1990s include the decline of price relativities, the stagnation of production due to the lack of demand and hence the incomplete use of the production potential of agriculture, the fall in the income from agricultural production; the limitation of subsidizing agriculture from the budget; and the increasing liabilities of farm households, as well as the considerable increase in the unemployment rate, the slower tempo of the overall economy and the deterioration of the economic situation in many areas.

An increase in farmers' income should be related to an increasing competitive capacity of farms rather than interventionism by the state, which on the one hand arouses the unjustified hope of rural people and on the other hand causes their frustrations. However, the state should endeavour to continue the increase of the importance of income from work on one's own account. This can be achieved by an appropriate macroeconomic policy that achieves multifunctional rural development. A situation where social services come first or second in the structure of households' income is unfavourable and proves that agriculture is in a bad condition.

Finally, it must also be stressed that in the 1980s and 1990s there were various trends concerning dual professions and gaining additional income outside agriculture and a different tempo of changes in individual countries of the Union, because there was still a relatively rapid decrease in the number of farms and people were migrating from the country to the city. But in the same period farmers' households noted an improvement in average disposable income in comparison with other social and occupational groups. However, there are still considerable differences between the EU countries as far as agricultural income and disposable income are concerned even if they were reduced during the period of the Common Agricultural Policy.

The Income Situation of Farmers in non-European Countries

In the USA in general, farm and non-farm household incomes are similar within the overall distribution. Mean incomes are similar for non-farm and farm households, though farm household income is more dispersed – larger shares of farm households have negative incomes and have incomes above $200,000. However, average wealth for farm households is substantially greater than for non-farm households, and is less dispersed. The farm business as a source of income has become increasingly less important to farm households, especially among farms with sales of less than $250,000 per year, which make up over 90% of all farms (Rural Households' Livelihood and Well-Being, 2007). Sources of income in the agricultural sector in the USA between 1969 and 2003 are shown in Fig. 2.6.

In the United States the average money income of farm households periodically exceeded that of all households, starting in the 1970s until the mid-1990s. Since then, the income of farm households has been consistently higher (Fig. 2.7). Average farm household income in 2003 was about $68,500, compared with $59,100 for the average non-farm household. Median income for farm households has also been roughly on a par with the median income of all United States households in recent years (Rural Households' Livelihood and Well-Being, 2007).

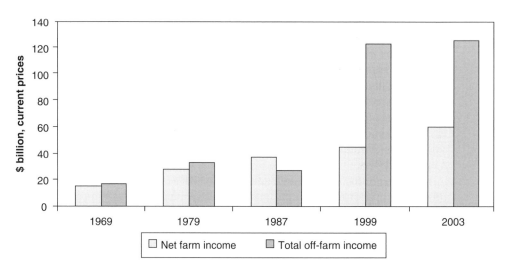

Fig. 2.6. Sources of income in the agriculture sector in the USA (1969–2003) (from Rural Households' Livelihood and Well-Being 2007, p. 336).

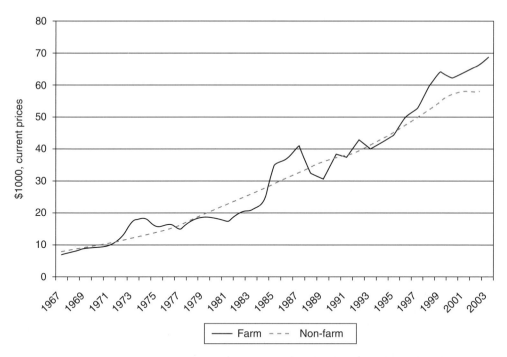

Fig. 2.7. Average income of farm and non-farm households in the USA in the period from 1967 to 2003 (in $1000, current prices) (redrawn from: Economic Research Service of US Department of Agriculture; Income of the agriculture household sector, 2001 report. Eurostat; Rural Households' Livelihood and Well-Being, 2007, p. 338).

In Australia in 2001, households that contained at least one person whose main income came from agriculture had a mean income of about 90% of those households where no person was employed in agriculture (Table 2.2). However, average incomes vary widely depending on the degree to which income from agriculture contributes to the total income of the agricultural household. If income from agriculture contributes less than one-quarter of total income, the mean income of the agricultural household is only 87% of that of non-farm households. Where income from agriculture constitutes between one-quarter and one-half of total income, the total income of the agricultural household jumps to 114% of non-agriculture households. If agriculture income accounts for between one-half and three-quarters of total income the agricultural household income drops to 97% of non-agricultural households. Where more than three-quarters of income comes from agriculture, the income falls to 76% of the non-farm income (Fig. 2.8).

It is evident from the data that farmer households as a group continuously earned a lower income than the other groups. A comparison of the monthly disposable income of farmer households with that of other households is to the farmers' disadvantage. The income per head in a farmer household was lower than in the other social groups. Only pensioners' and annuitants' households and the households of persons with a non-profit source of income recorded a lower income

Table 2.2. Income of agricultural and other households in Australia, by contribution of agricultural income to total income in 2001, $A. (From Rural Households' Livelihood and Well-Being, 2007).

Specification	Estimated number of households	Mean agricultural income ($A per week)[a]	Mean total income ($A per week)	Agricultural income as % of total income
Household contains at least one person whose main job is in the agriculture industry, where the contribution of agricultural income to total income is:				
Less then 25%	88,704	77	849	9.0
25% to less than 50%	40,415	424	1,110	38.2
50% to less than 75%	58,635	594	945	62.8
75% or more	78,201	673	743	90.5
Total	265,955	419	879	47.7
Household contains no person employed in the agriculture industry	7,048,965	–	975	–
Total	7,314,920	15	972	–

[a]Income from wages and salaries from main job plus own unincorporated business income where industry of main job is agriculture.

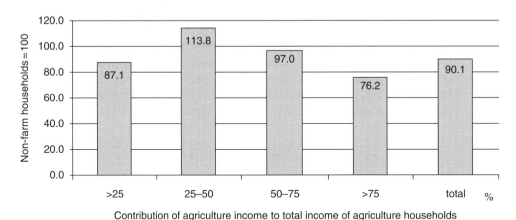

Contribution of agriculture income to total income of agriculture households

Fig. 2.8. Income of agriculture households compared with non-agriculture households (= 100) for different levels of contribution of income from agriculture in Australia, 2001 (from: Australian Bureau of Statistics, Survey of income and housing costs, 2000–2001; Rural Households' Livelihood and Well-Being, 2007, p. 392).

than the farmers. Farmers' anxieties, which have been observed in Poland since 1989, are mainly the consequence of their poverty. One may suppose that these anxieties will not disappear soon. In the 20th century the farmers' situation around the world developed differently depending on the region of the world or country.

In some countries rural people became even poorer, whereas in others their material situation improved. In the contemporary world there are attempts to solve the problem by giving rural people opportunities for higher earnings, which has become part of the economic and social policies of states.

Non-agricultural Sources of Farmer Income

In view of the economic situation of people related to agriculture, traditional family farms may be forced to combine income from several sources. The income comes from non-agricultural work, e.g. from running small service, trade or craft enterprises, or it has a non-profit character and comes from other transfers into agriculture, such as pensions and annuities. Underutilized labour and capital resources, especially in small-area farms, undoubtedly favour the search for extra sources of income and taking on extra activities by the people related with agriculture.

Sources of income for agricultural households can be divided into two main groups:

- Farm sources – activities related to agriculture;
- Off-farm sources, e.g. profit work outside one's own household (wages, salaries, income from other activities), non-profit sources of income (e.g. retirement or social benefits, interest and dividends) (Fig. 2.9).

The changes in the structure of rural people and in the structure of their income, depending on the country and the region, began to take place in the 1960s or at the beginning of the 1970s, with the increasing industrialization of agriculture. This is when typical peasants began to lose their dominant position among rural people, while agriculture began to lose its dominant position among the sources of income. Dual professions (or even multiple professions) in individual farms, especially those of small area, increased as the economy became increasingly industrialized. Farmers can be increasingly regarded as 'rural entrepreneurs'

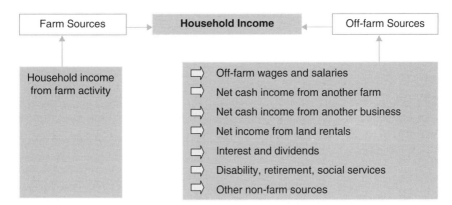

Fig. 2.9. Measurement of household income from farm and off-farm sources (based on Rural Households' Livelihood and Well-Being, 2007, p. 330).

who produce a whole range of goods in addition to agricultural commodities, provide services and combine a range of skills in the technical, financial and commercial fields (Rural Developments, 1997).

In such socio-economic conditions, first a conception and then a policy of multifunctional (multidirectional) rural development were engendered, which assumed supporting and developing non-agricultural initiatives in rural areas. However, it must be noticed that 'multifunctional rural development' and 'multifunctional development of rural areas' are not equivalent, though relatively close to each other. In fact, multifunctional development of rural areas may take a course that will result in the destruction of the rural structure. An example of this could be the location of international supermarkets or petrol stations by through roads in rural areas or chaotically emerging recreation sites that form dead enclaves in the country. Non-agricultural business activity is one of the main elements of multifunctional rural development. Agricultural farms develop by changing the profile of their traditional production of agricultural raw materials with their existing resources of buildings, people, capital and land. This enabled new non-agricultural action to be introduced in sectors such as: small-scale retail activity, rural tourism and agritourism, food and beverage and accommodation activity, agricultural product processing, cultivation of special plants (e.g. herbaceous plants and spices, fibre crops, plants to be processed for energy purposes, tobacco), craft and manufacture of souvenirs (pottery, sculpturing, manufacturing household objects of wicker, wood, folk painting, embroidery, etc.), building, transport services, communal services, landscape care (forest planting, biotope care, etc.) and other services. Therefore, it is important to determine the relation between the scale and tendencies in the sector and the existing potential of individual countries or even regions.

Urbanization and its Consequences for Agritourism

Urbanization is a continuous and permanent increase of the urban population at the expense of rural areas. It is an ongoing phenomenon around the world. As centuries pass urban people become more and more numerous relative to the number of rural people. Figure 2.10 shows the development of populations around the world during the last 60 years, as well as the forecast increase in the population to the year 2030. At present the world population is approximately 6.5 billion people. The proportions of people living in rural and urban areas are almost identical, the figures being 3.17 and 3.28 billion. Within the next 20 years the world population will increase to over 8.1 billion people, including 4.9 billion living in towns and cities. The rural population worldwide will reach its peak in 2010, followed by a gradual decline.

In the 1950s the rural population around the world was still more than 70% of the total. Over time, the share of rural population was decreasing and the share of urban population was increasing. In 2008 the urban population probably equalled the rural population and it will drop to 39% in 2030 (Fig. 2.11). This means that an increasing number of people worldwide will not have any contacts with the country and will not produce food. This might also result in a considerable increase in interest in these areas.

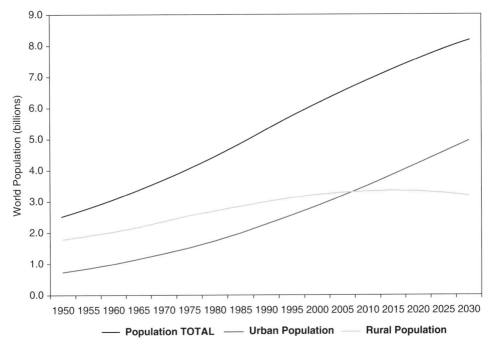

Fig. 2.10.   Data on the total world population, urban and rural populations in the years 1950–2008 and forecasts up to 2030 (based on FAOSTAT 2002).

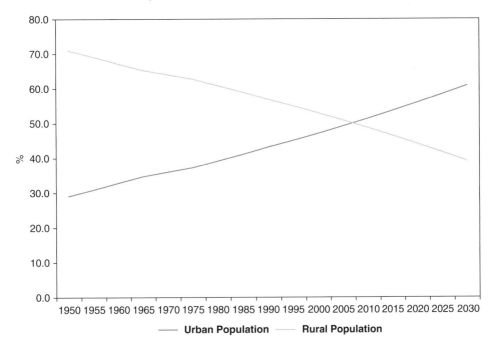

Fig. 2.11.   The percentage of urban (urbanization percentage) and rural population around the world (based on FAOSTAT 2002).

Figure 2.12 presents the development of the European population starting from 1950 and the forecast figures up to the year 2030. A rapid population growth in the early 1990s resulted in adjustment to statistical records after the collapse of the Soviet Union. The populations of the Baltic republics, Belarus, Ukraine and the republics in the Caucasus were included in the statistics of the European population, and not, as was the case before, in those of the Soviet Union. The number of people living in Europe has been decreasing systematically. At present there are 727.4 million Europeans, while in 2030 there will only be 685.4 million.

As can be seen in Fig. 2.13 in 2005 the rural population of all Europe was 26.7% whereas the urban population was 73.3%. The situation has been stable for some time, with the proportion hardly changing. In Europe the number of rural people began to decrease rapidly soon after the end of the Second World War, when the share of the urban population rose to more than 50%. However, it must be stressed that not all rural people deal with agriculture. It is worth noting that the percentage of agricultural people is decreasing slightly more rapidly than the share of rural people.

The term 'city' used previously is at present not sufficient to characterize the population of urban areas. Around administrative city limits people live in satellite settlements or towns. Huge areas are referred to as urban areas. The number of

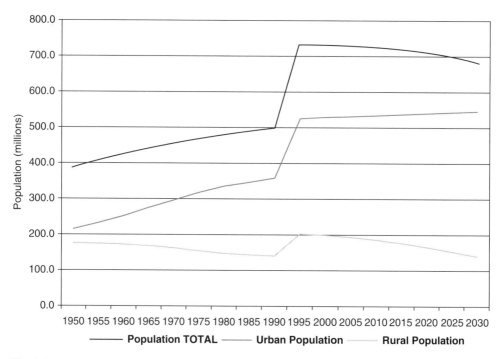

Fig. 2.12. The population of Europe, the urban and rural populations in the years 1950–2008 and forecasts up to 2030 (based on FAOSTAT 2002).

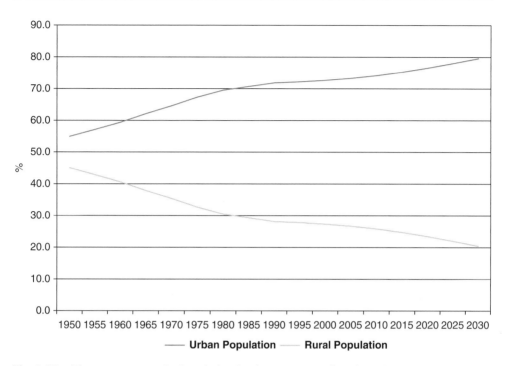

Fig. 2.13. The percentage of urban (urbanization percentage) and rural population in Europe (based on FAOSTAT, 2002).

people living in a city is generally much lower than the number of inhabitants of urban areas located around the city. Statistics give the population of a given city and its urban complex. For example, Paris has only 2.1 million inhabitants, whereas the metropolis has 10.6 million people. The population of Warsaw, the biggest Polish city, within its administrative limits of 2006 was 1,700,536, while that within the metropolitan area was approximately 2,600,000. There are two conditions that need to be met for a given area to be classified as an urban area. First, the population should be over 500,000. Second, urban areas are those in which the minimum population density is 400 persons per square kilometre (in Australia only 200). 'It is estimated that the urban areas with over 500,000 population account for 44.3 percent of the world's urban population as of 2002' (Demographia World Urban Areas: Population Projections 2007 and 2015, 2008).

The structure of cities is being transformed. Large urban areas are increasing at the expense of villages or towns. Table 2.3 compares the number of world urban areas in 1960 and 2007.

In the past a city with a population of 1 million was considered to be very big. Today cities of this size are nothing special. Figure 2.14 shows the cities whose population exceeded 1 million in 2006. A unique phenomenon of the second half of the 20th century was the rapid formation of megacities, whose population, in the case of Tokyo, reached as much as 35 million. In 1950 only two cities were over 10 million inhabitants, New York and London. London is

Table 2.3. The development of world urban areas with populations over 0.5 million in the 1960–2007 period. (Based on Statistical Yearbook 1965; Demographia. World Urban Areas (World Agglomerations): 2008. http://www.demographia.com/db-world-ua.pdf.)

Urban areas (millions)	Years		Growth multiplication factor in 1960–2000
	1960	2007	
Over 10	2	21	10.50
5–10	7	29	4.14
2–5	31	108	3.23
1–2	67	186	2.70
0.5–1	n.a.	363	
Total	107	707	n.a.

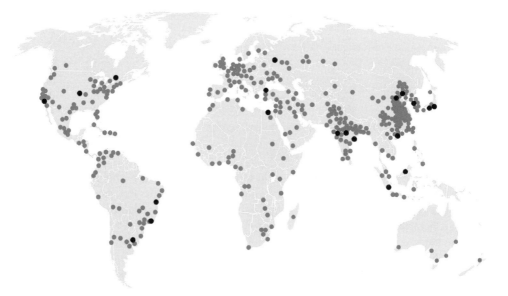

Fig. 2.14. The bubble map shows the global distribution of the top 400 cities with at least 1,000,000 inhabitants in 2006 (redrawn from: http://en.wikipedia.org/w/index.php?title=Image:2006megacities.PNG&).

unique as already, starting from 1939, its population was decreasing, so that at present it has only 8.3 million inhabitants. Today worldwide there are at least 21 megacities with a population over 10 million, namely one in Africa: Cairo, three in South America: Buenos Aires, Rio de Janeiro and São Paulo; two in North America: New York on the east coast and Los Angeles on the Pacific coast, one in Central America: Mexico City, 13 in Asia: Beijing, Calcutta, Delhi, Istanbul, Jakarta, Karachi, Manila, Mumbai, Osaka, Seoul-Incheon, Shanghai, Shenzhen and Tokyo; and two in Europe: Paris and Moscow. It is expected that by the year

Table 2.4. Urban areas of over 10 million inhabitants (data in million people). (From Demographia World Urban Areas: Population Projections 2007 and 2015, 2008.)

No.	Urban area — Cities of over 10 million	State	1950	1975	2007[a]	2020[a]	Estimated annual growth rate (%)
			2	5	21	29	
1	New York	United States	12.3	15.9	20.4	21.8	0.50
2	London	United Kingdom	12.1	n.a.	8.3	8.3	0.00
3	Tokyo–Yokohama	Japan		19.8	34.5	35.7	0.29
4	Shanghai	China		11.4	14.5	19.4	2.23
5	Mexico City	Mexico		11.2	18.4	20.5	0.85
6	São Paulo	Brazil		10.0	18.1	20.5	0.93
7	Seoul-Incheon	South Korea			20.1	21.7	0.60
8	Mumbai	India			19.4	25.7	2.20
9	Jakarta	Indonesia			19.3	28.3	2.98
10	Delhi	India			18.6	28.4	3.33
11	Manila	Philippines			17.3	21.5	1.68
12	Osaka–Kobe–Kyoto	Japan			17.3	17.5	0.08
13	Cairo	Egypt			16.0	19.7	1.61
14	Los Angeles	United States			15.4	18.6	1.50
15	Calcutta	India			14.6	18.0	1.65
16	Moscow	Russia			14.1	14.8	0.36
17	Buenos Aires	Argentina			13.5	15.2	0.92
18	Beijing	China			12.2	17.0	2.62
19	Shenzhen	China			11.8	18.9	3.65
20	Rio de Janeiro	Brazil			11.1	12.3	0.82
21	Istanbul	Turkey			11.0	13.5	1.57
22	Paris	France			10.6	11.8	0.83
23	Lagos	Nigeria				17.0	4.45
24	Karachi	Pakistan				13.9	3.19
25	Kinshasa	Congo (Dem. Rep.)				13.6	4.24
26	Ho Chi Minh City	Viet Nam				12.1	3.19
27	Dhaka	Bangladesh				12.0	3.71
28	Dongguan	China				10.3	3.65
29	Lahore	Pakistan				10.0	3.16
30	Chicago	United States				10.0	0.74

[a]by 2007 forecasted population

2020 this number will increase to include the next eight megacities. Table 2.4 presents the development of the populations in the biggest cities of the world.

Relationships between cities and villages are diverse. Towns and cities are a source of wealth for villages for several reasons:

- First, cities absorb huge amounts of food produced by the population working in agriculture.
- Second, cities supply means of production and services for agriculture and the rural population.

- Third, they become a workplace for a considerable proportion of the rural population.
- Fourth, cities supply the rural population with cultural and artistic goods and services.
- Fifth, city dwellers see rural areas as a place for recreation.

We need to realize that an increasing proportion of the world population, especially in urban areas, for practical reasons no longer have contacts with agriculture, the countryside and food production. However, contacts with nature, agriculture, animal breeding and rural areas are a natural and emotional human need. Agritourism is developing not only in order to supply additional income to farmers, but primarily to compensate for these natural and emotional needs. There are concerns whether agriculture will be capable of meeting mass demand for agritourism, which is still a sleeping giant.

The development of cities has always been accompanied by people's migration from the country to the city. Figure 2.15 shows the flow of people from the country to the cities. The rectangle on the left marks the urban population and that on the right the rural population. Part of the rural population emigrates from rural areas and settles permanently in the city. Such people are called migrants. During the post-Second World War period the migration of urban population to the country has been insignificant.

There is also a group of commuters travelling to cities daily in order to earn money, get an education, etc. Usually in the evening these people return to their

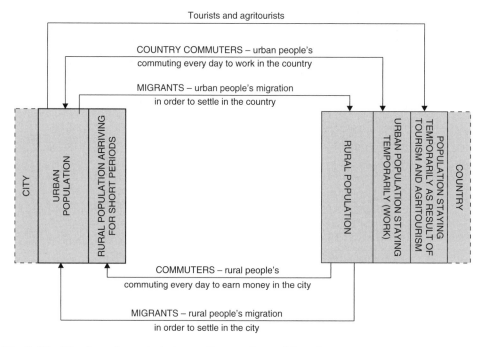

Fig. 2.15. The flow of people between the country and the city.

places of residence. Generally this consists of the rural population commuting to cities. However, an opposite phenomenon may also be observed – city dwellers commuting to villages to work (country commuters). However, this flow of urban people to the country is much smaller than in the opposite direction.

There is also one more component of people movement, related to tourism and agritourism: from urban areas to rural ones. This relocation is for a limited period of time: for a day or two, a week, a month, etc. Because such a flow of people provides income for agritourist farms, it is also an object of interest to agritourism.

If people moving from villages to cities find employment and housing there, i.e. no negative economic or social phenomena occur, then such migration may be considered advantageous. However, in many countries, for many years, rural people have left the country and moved to cities, forming agglomerations with huge zones of misery, poverty and slums (Fig. 2.16), where young people have no prospects for the future. For many people, abandoning rural areas is equivalent to transferring their misery from rural to urban areas. This development of cities is not an attractive phenomenon.

Social development in developed countries consists in the fact that apart from the growth of population in big city agglomerations, the state tries to keep the population in the country by giving them a chance to earn in non-agricultural sectors. Thus, people's migration from the country to cities is a positive

Fig. 2.16. Uncontrolled migration of rural population leads to the formation of misery zones, with poor people and slums around big cities (Vadodara, India) (photo by M. Sznajder).

03 11 2001   12:15

Fig. 2.17.   A clean, neat tourist city is a sign of urban people's prosperity (photo by M. Sznajder).

phenomenon as long as it does not cause negative economic and social phenomena. If the people leaving the country and settling in cities find employment there, the migrations can be evaluated as positive (Fig. 2.17). Unfortunately poor rural people, by moving to the city, very often cause even bigger social problems, related to safety, drug addiction or prostitution. If the city development is unsuccessful, this phenomenon is socially harmful.

In this context many countries try to create employment opportunities outside agriculture for rural people. There are two methods of creating such opportunities. The first method is facilitating the search for employment in the city; simultaneously forming the infrastructure in such a manner that rural people can commute to the cities in the morning and return home in the evening. A condition for facilitating this approach is developing the transport system by constructing fast roads and a system of commuter railways. A superficial analysis of the road and rail network in Holland, Belgium, France or Germany shows that people's migration in these countries is very easy. Commuting over the distance of 100 kilometres in the conditions of good transport infrastructure does not pose a barrier for this method of activating rural people. Transport lets people live in the country and work in the city.

The other method is activation of rural areas. This consists in giving rural people a chance to gain extra income in the country (Fig. 2.18). In rural areas various types of enterprises and services are established. Rural areas are treated as a source of a low-wage labour force, where investors eagerly invest their money. Some countries, especially the countries of the European Union, run a special policy whose goal is to keep rural people in the country by giving them a chance to earn outside agriculture. This programme is known under the name of

Fig. 2.18.   A town in Tuscany, Italy, gaining income from agritourists (photo by B. Moskalik).

multifunctional rural development. It comprises various types of subsidies for rural people, e.g. a grant for conservation of the environment, which function in order to encourage farmers to stay in less favoured areas. Another way to keep rural people in the country is the manufacture and service activation of these areas, which consists in establishing various service companies in the country, reducing taxes for them and facilitating their functioning in other ways (Fig. 2.19).

One of the methods of activation of rural areas is agritourism. The role of agritourism consists not only in the creation of conditions for recreation for urban people in rural areas but also in the redistribution of income from urban to rural people.

Rural people try to increase their income by taking on various non-agricultural jobs in transport, building, trade or agritourism. According to the General Census of Agriculture of 1966, 8.1% of farms in Poland dealt with a non-agricultural activity. They were involved in building (0.8%), transport and storage (0.7%), trade (3.1%), agricultural services (0.2%), manufacture (1.2%), agritourism – hotel business and renting rooms (0.2%), forestry (0.5%), fishing and fisheries (0.4%) and other services (0.4%).

The Census of Agriculture of 2002 showed a considerable – 46% – rise in the number of farms whose user or adults sharing the household with the user ran a non-agricultural business (Report on the Results of the National Agricultural Census 2002, 2003). In 2002 12.4% of the total number of farms ran a non-agricultural business on their own account. More than half of the people connected with farms (3860.3 thousand, i.e. 51.8% of the total number) earned a living or were supported from sources other than the income gained from work

Fig. 2.19. Dzierżoniów, a small town serving rural people in Poland (photo by M. Sznajder).

on a farm. Starting non-agricultural activities by a growing number of farms, along with the persisting poor economic situation in many agricultural markets, was related to the search for new sources of income by the agricultural population. Similarly to 1996, the farm users who ran a non-agricultural business usually indicated activities related to trade (wholesale and retail), food processing, building and transport.

The non-agricultural activation of rural areas and technological progress in agriculture cause a constant decrease in the number of people earning a living from farming. In 2001 on a world scale, the number of agricultural population amounted to 41.9%, while in Europe it was 8.2% (18.5% in Poland).

Demographers and politicians realize what will happen if non-agricultural activation of rural areas does not take place, especially where 'agrarian overpopulation' can be observed. The answer is relatively obvious. In those parts of the country, people will migrate to cities. Not surprisingly, in many countries worldwide there are already depopulated rural areas.

Redistribution of Income from the City to the Country

Rural people are usually poorer than part of the city population. Therefore, redistribution of financial resources from cities to the country and increasing the possibility of rural people generating income are important goals of social policy. All types of migration to cities and from cities to the country result in a circulation

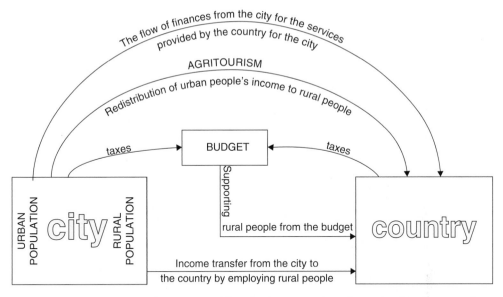

Fig. 2.20. Redistribution of income enabling the increase of rural people with allowance for agritourism.

of finances between the rural and urban population. The goal is to increase the flow of finances transferred from the city to the country. Figure 2.20 shows redistribution of income between the country and the city.

Redistributional activity by the state is often initiated to enhance social cohesion. Redistribution of income is done through many channels. Ideally, it develops on the basis of market rules. If rural people arrive at work in the city every day, the earned money is transferred to the country. The transfer of finances in the opposite direction, i.e. from the country to the city, also exists, though it is small. The migration from the country to the city improves the situation of the people remaining in the country. The finances that are generally available to rural people are constant, but, owing to the decreasing number of people who use them, those who stay in the country gain more income. The people migrating from the country to the city increase the per capita income of the people who stay in the country.

Another way to increase rural people's income is by the state budget. By taxes urban and rural population generate a budget. Aid programmes for rural people paid from the budget, e.g. direct subsidies, goal programmes, etc., cause redistribution of money from the city to the country. Budget redistribution is not very encouraging for the urban part of society and it is likely to cause conflict.

Another channel of redistribution of finances results from the fact that the country generates products and services that are sold to urban people. Agritourism is becoming an important element of redistribution. It is predicted that the importance of this method of redistribution will increase. Agritourism is an effective and arguably fair method of transferring finances from the city to the country. By agritourist activity the rural community has a chance to grow richer. When travelling around a country it is easy to distinguish between the wealth of rural people gaining profit from tourism and those who do not have this opportunity. In tourist resorts houses are renovated and homesteads look neat.

The importance of agritourism for rural areas consists in the fact that by being involved in this type of activity farms are capable of generating higher income. The finances that reach agritourist entities, as a result of another redistribution caused by the employment of people attending to agritourists, reach wider circles of the rural community. In the literature this phenomenon is defined as the multiplier effect. The concept of the multiplier effect of tourism was introduced in the 1970s by Albert Schmidt in his work entitled *Fremdenverkehr, Multiplikator und Zahlungsbilanz* (1970/71). Denman (1993) calculated that in the United Kingdom every pound spent by a tourist on accommodation in rural areas brings a further 66 pence spent on various types of goods and services within the rural economy in its broad sense. It is possible to gain a satisfying income in the country without moving to urban areas or commuting to the city every day. The occupational activation of rural people proves to be a cheaper method of social development of the population than rural people's migration to cities.

Agritourism has significant importance for urban communities. Urban people need relaxation, which can be provided by traditional tourist structures. However, more and more frequently such people want to relax in rural areas. Agritourism gives the possibility of alternative use of holiday and weekend time. It has an increasing importance for the economies of both highly developed countries and least developed countries. In highly developed countries the urban population relaxes on farms, which causes a redistribution of income from the city to the country. In poor countries foreign tourists who want to familiarize themselves with the production or lifestyle on farms can pay for this. Redistribution of finances from richer to poorer countries takes place.

Military strategists, who deal with minimization of losses among urban people in consequence of wars or other disasters in cities, recommend that every urban family should have acquaintances in rural areas. Agritourism enables the development of such relationships between rural and urban people, which may be valuable in such crisis situations.

Agritourism and the Sale of Agricultural Products

For many farmers agritourism is one possible sale channel for products and services, especially products manufactured in small quantities, which are not important from the point of view of supermarkets. As it turns out, the demand for niche products, i.e. usually a local product, which is not a subject of interest to big manufacturing companies, is sufficient to provide a means of living for many producers. It is only necessary to find the necessary sales channels. Farms dealing with direct sales usually form a sales outlet on their territory. Making tourists interested in these products and attracting them to the farms in order to purchase the products are possible and they are included in the development of tourist enterprises in the country. In the USA there is an association named the North American Farmers' Direct Marketing Association (NAFDMA), whose aim is to indicate the methods of

Fig. 2.21. Agritourism shop and restaurant offering a wide range of cheeses: Dairyland shop, Hawera, New Zealand (photo by L. Przezbórska).

increasing effectiveness of direct sales, which include the use of agritourism. Figure 2.21 shows a farmer's shop offering the farm's own products.

The available data do not allow a full description of the role of agritourism in the national economy. What is particularly necessary is a presentation of: the contribution of agritourism to the GNP, the number of farms dealing with agritourism, the number of rural people professionally involved in agritourism, the number of urban people using agritourist services and the proportion between the income from agritourism and the income from agriculture. So far it has been impossible to obtain accurate data on the economic importance of agritourism according to the aforementioned indices for Poland and the European Union countries because of no appropriately separated statistics. Agritourist activity in the statistics of the European Union countries is included in tourist activity despite the fact that from the point of view of activation of farms it could be considered as a non-agricultural activity of farms.

Case Studies

1. Define the boundaries of a region (state, province, region, district, commune), as well as specific farms, which you will always refer to when considering case studies given in this book.

2. What does the rural population in the selected region do? How would you define the profitability, financial situation and standard of living of this population in relation to the urban population?

3. Using data from a statistical yearbook define the following for the population of the region you have selected: proportions of the rural and urban populations and the percentage of the rural population not working in agriculture.

4. List urbanized areas in your region together with their size. What are the links between city dwellers with the countryside and with agriculture on the one hand, and those of the rural population with urban areas?

3 Multifunctional Development of Rural Areas

The challenge of multifunctional rural development and tourism, including rural tourism and agritourism, which are an integral element of such development, is fairly frequently mentioned in the literature. However, the term 'multifunctional development' is understood differently in different regions of the world, depending on the extent of economic development of a country, the role of agriculture in the economy and employment and the rural population. Not only is development interpreted as economic growth and improvement in the economic situation of rural people, but it is also assumed that it is a balanced demographic, economic and social development. Such a broad definition of multifunctional development is particularly significant in countries where rural communities, especially farming communities, have a low share in the total population, e.g. in the United States, Canada, New Zealand or Australia. In general, rural development denotes the actions and initiatives taken to improve the standard of living in rural areas (non-urban neighbourhoods, countryside and remote villages). Agricultural activities may be important in this case, while other economic activities would relate to the primary sector, production of foodstuffs and raw materials, as well as the other sectors, the secondary (e.g. manufacturing foodstuffs) and especially the tertiary sector, i.e. services.

This chapter gives a broad macro- and microeconomic background to the challenge of multifunctional development. The chapter also discusses agritourism and rural tourism within the concept of multifunctional rural development, with special attention given to the conditions that influence or may influence the starting, functioning and development of farms dealing with agritourism and rural tourism.

Transformations in Agriculture and Rural Areas

The 20th century, especially its second half, brought enormous economic development to European countries and very serious adjustments in the economic structure. Agriculture, one of the oldest production activities of human beings, is no longer the dominant activity in many economically developed countries and its share in the economy is reduced in favour of other activities (Rural Developments, 1997). Agriculture, which for many years was treated as the only or main branch of economics in rural areas, is subject to certain universal trends, which include: an increased concentration of production and farms, a decreasing number of farms and declining percentage of people employed in the sector, a lower share of agriculture in the composition of the gross domestic product (GDP), the increasing efficiency of agriculture and also the relative decline in the demand for food and the farming products necessary to produce it. According to the Fisher Clark and Fourastié theory of economic development, technology, work efficiency and the level of real salaries influence the evolution of the general structure of employment: the increasingly advanced level of economic development causes a relative and then even an absolute fall in employment in primary sectors (i.e. agriculture, fishing, forestry and mining), first in favour of employment in manufacturing (secondary sector) and at later stages in favour of employment in services (tertiary sector) (Fisher, 1935; Fourastié, 1949). The consequence of these changes is the tightening of connections of agriculture with other sectors of the economy, an increased share of other sectors in the food value chain, changes in the production scale, etc., and eventually the formation of a complex called agribusiness and the emergence of the concept of multifunctional rural development. The term agribusiness was introduced into the scientific literature in 1957 by Davis and Goldberg, who used it for the first time in the work *A Concept of Agribusiness*. What they meant by agribusiness was 'a system of integrating farmers with supply, food processing and distribution entities, which allows effective control of all interdependent elements, from the farm to the supermarket and consumer' (Davis and Goldberg, 1957).

The beginnings of the concept of multifunctional rural development and the first work on their practical implementation could be found as early as the 17th century in North America. At the time, actions to develop rural areas were limited mainly to self-help projects for the inhabitants of American rural areas (Ragland, 1996). The real recognition of the need to improve the living conditions of rural inhabitants took place only at the beginning of the 20th century. At the time, many rural communities in America were in stagnation and were afflicted by social or economic crisis, similarly to certain regions of Europe. It was then that the United States Department of Agriculture (USDA) began to be more involved in actions for rural areas. One of these was creating a Land Grant University in each state, which by definition was to deal with all aspects of problems of rural areas.

As early as the 1960s in Europe, problems related to the concentration of non-farming business activity in cities began to be noticed. Another problem resulting from the situation was the degradation of smaller towns, though at the time the problem of rural development was marginal (Rural Developments, 1997). The rapid development of technology and industry, demographic changes, which included a rapid growth of population, the progressive urbanization of many European countries, depopulation of outskirts as a result of migration of rural people to cities and then also the processes of ageing of societies and the related economic changes increasingly influenced the traditional image of rural areas, country and agriculture (Whitby, 1994). Although the essential functions of rural areas still comprise activity in agriculture, forestry and fishing, more and more frequently the production and service activities are also listed, i.e. exogenous functions such as: industry, building and extra-local transport, residential construction, tourism and recreation, as well as education and health services. Among the supplementary functions of rural areas, the types of socio-economic activities most frequently listed involve production for the manufacturers' own needs or local needs, e.g. self-provision farming, craft and sometimes industry, and those activities that serve the local people, e.g. education or health care. In many countries the processes of functional transformations of rural areas led to a situation where the country is no longer the area where only farmers live and work. More and more frequently, people who do not have much in common with agriculture live there. The definitions of the country and rural areas that have been used to date not only lose their old sense, but in the future the only fully country-related and dynamically developing function will be recreation and tourism. The changes in rural and farming areas and the population residing there have led to a change in the structure of rural people's sources of income, where the farming activity began to play a less and less important role. However, one must remember that agriculture as a primary sector is still very important and farmers are a very important social group. The farming sector serves rural communities. Its role is not only to produce food but also to guarantee the survival of the countryside as a place to live (residential function), work (in various activities) and visit (tourism and recreation).

Concepts and Definitions

Multifunctional rural development is a very broad term, which is variously interpreted. It is commonly considered to be rural activation and diversification of business activity so that the future of rural people is connected not only with agriculture but also with the branches of the economy that are alternative to agriculture. It is particularly concerned with creating new opportunities for employment and overcoming unemployment, searching for alternative sources of income in professions related to the agricultural environment and the diversification of profit-generating activities using rural resources. Usually this concept appears in the context of declining agriculture within rural areas and the decreasing role of agriculture in the national economy. The strategy of multidirectional rural development consists in diversification of the rural economy, thus abandoning monofunctionality, which usually involves the production of agricultural commodities or raw materials.

The model of monofunctional rural development and locating non-agricultural activities only in towns was perceived to be unfavourable by many due to the increasing urban housing crisis, the depopulation of certain rural areas and the deformation of demographic structures. Over the years, high explicit and implicit unemployment resulted, which is difficult to eradicate.

Multifunctional development involves introducing an increasing number of new non-agricultural functions – production, trade and services – into rural space. Thus the country ceases to be an area inhabited by traditional farmers where only farming materials are produced, but becomes an integrated part of the national economy, a place inhabited by people related to both agriculture and non-agricultural branches of life. Employing rural people in branches alternative to agriculture is attractive to many people, gives people greater employment choice and contributes to economic diversification. This not only contributes to increasing people's income but also increases the attractiveness of the country as a place to live and work in. So multifunctional rural development exists in parallel with the development of agricultural production activity, non-farming business activity related to agriculture and non-farming business activity not related to agriculture. The above-mentioned groups of activities may activate rural inhabitants economically. In the 1990s Jenkins *et al.* (1998) called these kinds of changes in rural areas 'industrialization'.

Elements of Multifunctional Development

The basic elements of multifunctional rural development include (Kamiński, 1995):

- Agricultural production activity (farming, i.e. plant and animal production).
- Non-agricultural activity directly related to agriculture:
  ○ provision of materials and means of production;
  ○ production services (e.g. mechanization, redecoration and construction, plant protection, etc.);
  ○ purchase, storage, transport and trade in farming products.
- Non-farming activity, indirectly related or completely unrelated to agriculture, including:
  ○ rural tourism and agritourism (organization of recreation and holidays, food and beverage services, accommodation services);
  ○ forest economy;
  ○ landscape care and environment protection.
- All other production and service activities unrelated to agriculture (public services and activities, small and medium enterprises (SMEs) representing various businesses).

The processes of multifunctional rural development may run in two directions: as an external inflow of ideas, capital, conceptions and organizational solutions and as the development of local enterprise and economic activity of the community of rural inhabitants.

Multifunctional rural development is subject to the very strong influence of space conditions, including the following:

- Demographic factors (e.g. population and its structure, density, migration, dual professions, unemployment, qualifications and education, etc.);
- Natural (earth resources, soil quality, climate, terrain sculpture, forest density, scenic values, etc.);
- Capital;

- Infrastructural (mainly technical and social infrastructure);
- Others (especially those related with the state's regional policy, region location, agrarian and property structure).

The factors conditioning rural development can be divided into two groups: endogenous and exogenous factors. The endogenous conditions include work resources, rural production potential, technical infrastructure of rural areas and know-how level, whereas the exogenous factors, i.e. those affecting rural development, include the macroeconomic efficiency of the national economy and agriculture (including the GNP level, the contribution of agriculture to the GNP, the productivity of the labour force in the agricultural sector, the effectiveness of the material outlay of the agricultural sector in the make-up of the GNP and the macroeconomic and agrarian policy of the region and country, i.e. legal and political conditions).

Multifunctional development depends on economic conditions, which include farming effectiveness, work efficiency, production increase, improvement in the area structure of farms, creating new employment opportunities in small towns and villages (including non-farming employment), counteracting people's migration to cities and depopulation of regions, new forms of economic activity and the so-called 'multiplier effect' of the actions.

Rural development is determined by the location of the commune and the so-called 'location benefit', the historical structure of the economy, the strength and character of agriculture in the development of the micro-region, the condition of the infrastructure, the demographic situation in the community, including the unemployment rate, social, job and property relations and capital resources, the efficiency of rural institutions (including local governments), social attitudes (predisposition to enterprise) and the type of local community and its sociocultural qualities. What can also favour the development of alternative sectors in rural areas is the structure of farms and the unused labour and capital resources (including housing resources). All the aforementioned factors have a stimulating or inhibiting effect on multifunctional rural development, but they also influence the decisions to start tourist enterprises and their functioning, which also constitute an element of multifunctional development.

The Origin and Development of Multifunctional Rural Development Concepts

In Europe the first serious proposals to develop rural areas appeared along with the establishment of the European Economic Community, though a uniform concept of rural development policy originated only in the mid-1980s. Then, as part of the Common Agricultural Policy (CAP), there was a proposal to abandon the philosophy of agricultural production maximization in favour of a rural development policy by intensification of production, systems of land-idling, the introduction of new, non-farming functions in rural areas, the promotion of employment forms alternative to farming and supporting structural policy.

By agreeing on the Common European Policy in 1957, the signatories of the Treaty of Rome committed themselves to common policies, including rural development policy. It is made up of three elements, agricultural, structural and regional policies, and the most important objective is increasing the economic and social uniformity of the Union by supporting the less developed regions and reducing disproportions in the development standard, thus bridging the gap between the living standards in different regions. However, over the years that have passed since the establishment of the Union, both the importance and the expenses of the realization of these objectives have risen considerably. The policy of the European Union is based on granting subsidies and coordinating the structural policy of the member states in order to support all-round and harmonious development, steady and balanced economic growth, reinforcing its economic and social integrity (according to Articles 2 and 130A of the Rome Treaty). In order to achieve this, the member states endeavour to level the differences between them by means of structural funds. Thus the structural funds are the main source of financing the Union's rural development policy. Because tourism, including agritourism, plays a particular role in regional development, being a sector of the economy that may help to develop the poorer regions with tourist potential, the European Community gives the possibility to receive financial support to develop tourism in rural areas within the funds from the European Regional Development Fund and the European Agricultural Guidance and Guarantee Fund.

In view of the strong development of structural policy and its increasingly closer connections with environment protection and other directions of the development of rural regions, there was a proposal to transform the Common Agricultural Policy into the Common Agricultural and Rural Policy for Europe (CARPE). A CARPE, of course, will include farmers, agriculture and food-related activity. Just as rural is not synonymous with agricultural, rural development is much wider than agricultural development. Thus a common rural policy is not just concerned with farmers and agriculture. Agriculture was at one time the dominant industry and employer, but this is no longer true. These will always be very important parts of rural areas and policy, but the policy must go considerably beyond them (Buckwell Report, 1997).

Since the time of the McSharry reforms, we have been able to speak of gradual transformation of the CAP into the Common Rural Policy (CRP) (Buckwell Report, 1997). The best proof of this fact is the increasing share of funds earmarked to realize this policy. The expenses on structural policy make up about 40% of the budget and together with the expenses on agriculture are more than two-thirds of the total budget of the Union.

The consolidation of the concept of multifunctional rural development policy took place in the 1990s. It was stated in the Single European Act that a balanced multifunctional development resulting from the rules of eco-development is an opportunity for rural areas. In a general sense it is identified with balanced development and it is defined as 'socio-economic development, where in order to balance the opportunities of access to the environment for individual communities or their members, of both the present and future generations, there is a process of integration of political, economic and social actions with the preservation of natural balance and permanence of basic natural processes' (Strategy of eco-development in a rural commune, 1999). The term sustainable development has occurred in the literature since the late 1980s and is ambiguous.

Usually it is interpreted as eco-development, which means development in agreement with the environment, even at the expense of limiting socio-economic development (the definition preferred by ecologists), as balanced development (which is closer to the term sustainable eco-development), stable development, which means balanced and stable (i.e. systematic, steady, constant, uninterrupted, continuous) or possibly as permanent development (which seems to be closer to the terms durable or lasting development) (Halamska, 2001). In the context of the definition of balanced development, each sector of the national economy can be considered, including the sectors related to rural areas, which are mainly agriculture, forestry, small industry and tourism. However, supporting different functions is based on the assumption that those production functions are important, but not always of primary importance. So the balance pertains to the commercial, social, ecological and also spatial and cultural spheres. The essence of this conception boils down to each region finding its own role and position, with the best possible use of local conditions and triggering the activity of territorial self-governments, local authorities and the rural communities themselves.

The EEC Commission document *The Future of Rural Society* of 1988 could be considered the next stage in the forming of a contemporary development model of European rural areas (*Future of Rural Society*, 1988). It emphasized the decreasing role of farming as an employment sphere and source of increase of the regional product. Simultaneously, more attention was given to the whole of rural areas, which were divided into dynamically developing rural areas near towns and tourist resorts, declining areas, whose development depends on the diversification and stimulation of the local economy, and depopulated areas or those in the course of depopulation, which require special support in order to retain at least the minimum population.

Other important steps taken on the way to forming the policy for developing EU rural areas were developing the *European Charter for Rural Areas* (1996) *and the European Conference on Rural Development – Prospect for Future* (1996). The *European Charter for Rural Areas*, initially accepted on behalf of the countries of the Council of Europe in 1996, is a document defining the new range of tasks enabling balanced and harmonious development of rural regions in Europe, making a foundation for permanent resource management by initiating new activities in the sectors of agriculture, forestry, fish breeding and fishing, including nature and landscape protection and participation in agritourism and recreation. The *European Charter for Rural Areas* distinguished the following directions of action in the rural development policy: the tourism development policy in rural areas, with special attention given to agritourism, the policy of multifunctional country and agriculture development, the employment development policy with special attention given to non-farming employment and the policy of people's income increase. All these directions are closely interrelated and complement one another. The Charter assumed diversified management of natural resources in rural areas, balancing economic, sociocultural and ecological functions and protection of the rural and cultural heritage. What is important is that the Charter also emphasizes that 'the parties should take necessary legal, tax and administrative action in order to develop tourism in rural areas with particular respect to agritourism. In order to achieve this it is necessary to support various forms of holidaymaking in the country and take initiatives which could encourage the farmer to receive tourists on the farm.'

Towards the end of 1996, the participants of a conference in Cork, Ireland, accepted a document, referred to as the Cork Declaration, entitled *A Living Countryside*, which included a programme concerning the formulation of the new policy of support of rural areas in the member states of the European Union. The programme comprised the essential assumptions of the policy of rural development. These included complex, integrated and balanced approaches to making an interdisciplinary policy of rural development in reference to both agriculture and diversified business activity, natural resources management, environment protection and support of cultural development, tourism and recreation, diversification of economic and social activity in the country and the durability of development, i.e. sustaining the unique character of rural landscape.

More changes concerning the policy of rural development are the consequence of a set of documents entitled *Agenda 2000, For a Stronger and Wider Union* (1997), which was prepared and published by the European Commission in 1997 and which defined the Union's financial and budget actions for the years 2000–2006. Ensuring a fair living standard for rural people and the stability of farming income, diversification of income, creating supplementary or alternative employment opportunities and sources of income for farmers and their families and the development of balanced farming remained the key objectives of the CRP. Environment protection, which includes the role of agricultural producers in natural resources management and landscape protection, also became more important. Thus, agriculture ceased to be treated as a key sector, at least as far as employment in rural areas of the EU countries is concerned, and is undergoing further transformation into a multifunctional sector. The types of activities that are most frequently mentioned in the context of diversification of the rural economy include not only public institutions, bank services, small and medium companies, telecommunication and information technologies but now also rural tourism.

The first legal acts of the Communities on tourism come from later periods than the foundation Treaties of the European Communities. They concerned the freedom of providing certain tourist services. However, the first regulations concerning the tourist policy come from 1984. They were the Act of the Council of 10 April 1984 on the Community's policy on tourism (*Official Journal of the European Communities*, 1984) and the document *The Community's Policy in Tourism. Initial Directives* (*Official Journal of the European Communities, 1984*). In the 1990s the documents and multifunctional rural development programmes published by the European Union showed a clear attitude of the Community to tourism in rural areas, and agritourism was formed. It is expressed in the statement, 'Rural tourism constitutes a fundamental element of the rural development policy, a very important and complementary component of the development strategy and plays a special role in country planning' (Opinion of the Committee, 1995). In many countries, such as Austria, Germany, Ireland and the United Kingdom, rural tourism is considered to be a very important factor stimulating the local economy, and to emphasize its importance it is called 'an industry of the future'. The policy of rural development and thus the policy of the development of agritourism must be compatible with the rules set by the European Union.

What is understood by the term multifunctional development is also active environment protection, landscape care and preserving the local culture and traditions. Otherwise the rural areas will lose their attractiveness and thus their economic potential. Hence agritourism and rural tourism are tools for the realization of many of the strategic goals of the community's policy, such as increasing employment, even part-time or seasonal, searching for alternative sources of income for farmers, activation of economically neglected regions and balanced (harmonized) development. Lasting, balanced and environment-friendly rural tourism enables the overcoming of certain problems of the areas located in unfavourable conditions and levelling the differences between the regions (Fig. 3.1). The multitude of functions ensures that, on the one hand, it is an object of interest to many sectors, not only tourism but also agriculture, regional policy, environment protection and small and medium business development, while, on the other hand, it also gives the possibility to support its development from various community funds. It is thought that it may contribute to the economic and social progress and diversification of rural areas.

Rural Europe, which 'extends across regions, landscapes of natural countryside, farmland, forest, villages, small towns, pockets of industrialization and regional centres' and comprises 'a diverse and complex economic and social fabric: farms, small shops and business, commerce and services, small and medium-sized industries', has changed very much over the last century (Rural Developments, 1997, p. 6). In the majority of European countries, changes in agricultural practice and policy due to intensification of production and modernization of farms have created unemployment in rural areas, falling agricultural income and the economic marginalization of small farms. Lack of employment opportunities led to outmigration to urban areas and depopulation of rural areas, while lower numbers of the rural population and the subsequent reduced demand for services led to a decline of rural services (Fig. 3.1; Yeoman, 2000). However, some European rural regions are the most dynamic economically. In some areas, there has even been a process of repopulation of rural areas as a result of a reverse migration trend of urban dwellers moving to the countryside (Yeoman, 2000) and considerable success in generating a higher level of new employment opportunities than the national economies as a whole, e.g. Tyrol in Austria, Almeria, Murcia, Valencia and Sevilla provinces in Spain or the Alps and the Atlantic Arc departments in France. 'The creation of rural employment resulted from a specifically territorial dynamic' and several factors influencing it, e.g. a sense of regional identity and social cohesion, an entrepreneurial climate, a good educational level and an attractive natural and cultural environment (Rural Developments, 1997, p. 16).

Within the context of rural change, rural tourism, agritourism and recreation in rural areas evolved due to an increase in the supply of opportunities (Yeoman, 2000) and will certainly see further growth over the coming years. At the same time, employment related to these activities on- and off-farm has become an important source of income in many rural areas, and many rural areas, especially in Europe, already benefit from tourism development (Travel and Tourism: a Job Creator for Rural Economies, 2000). The opportunities for rural tourism have been created by the need for more diverse rural economies. The context of tourism development within rural areas is summarized in Fig. 3.1 (Yeoman, 2000).

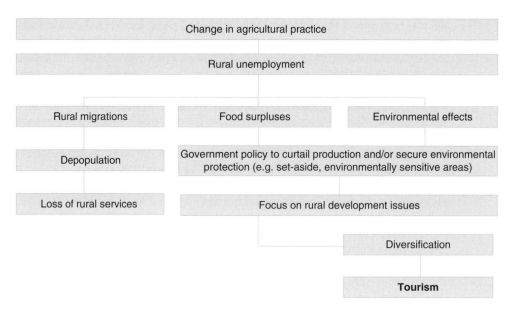

Fig. 3.1. The context of rural tourism (based on Yeoman, 2000).

The Community's Policy in Tourism: Initial Directives, submitted to the Council by the European Commission, confirms the separation of the Union's tourist policy and delineates the priorities for the Community's actions, which include retaining the architectural heritage in less developed regions and promotion of social, cultural and rural tourism (*Official Journal of the European Communities*, 1984). In turn, the Council's decision of 13 June 1992 delineates the Community's plan of actions aimed at the development of tourism, which include cultural, social, rural and youth tourism and connections between tourism and the environment (*Official Journal of the European Communities*, 1992). In order to implement the programme, training activities are organized (trips, seminars, exchange of experts, etc.) and action is taken in order to improve the information flow and access to assistance systems for people working in the rural environment and to encourage improvement in the quality of tourist services and support the enterprises facilitating access to the rural tourism market.

Following the fundamental reform of the first pillar of the Common Agricultural Policy in 2003 and 2004, and fundamental reform of rural development policy for the period 2007 to 2013, rural development is playing an increasingly important role in helping rural areas to meet the economic, social and environmental challenges of the 21st century (Rural Development Policy 2007–2013, 2007). 'The reforms have brought the Leader Community Initiative into mainstream programming and also have made an important step in simplification by bringing rural development under a single funding and programming framework' (New Perspectives for EU Rural Development, 2006).

As over half of the European areas are maintained by farmers (New Perspectives for EU Rural Development, 2006) and rural areas make up 90% of the territory of the enlarged EU (27), the new legal framework points more clearly to the direction of growth and creating jobs in rural areas (in line with the Lisbon Strategy) and improving sustainability (in line with the Göteborg sustainability goals) (Rural Development Policy 2007–2013, 2007). The European Union's rural development policy evolved from a policy dealing with the structural problems of the farm sector to a policy addressing the multiple roles of farming in society and, in particular, challenges faced in its wider rural context. It is very important that the development of rural areas is no longer perceived only as a base for agriculture alone, but diversification within and beyond the agricultural sector must be seen as unavoidable in order to promote viable and sustainable rural communities (New Perspectives for EU Rural Development, 2006).

Thus rural tourism and agritourism have become matters of European Union interest and policy. The challenge of forming policies for tourism and rural development has been enormous. However, it is worth noting that the transformation of the Common Agricultural Policy into the Common Agricultural and Rural Policy was in progress almost from the very beginning of the Union, starting from Mansholt's Plan of 1968, through the Green Paper of 1985, McSharry's reforms of 1991–1992, Agenda 2000, the 2003/4 CAP reform up to the fundamental reform of rural development policy for 2007–2013, as well as the other initiatives mentioned earlier. Nevertheless, the experience gained by the countries of Europe shows that the process of multifunctional rural development can be a significant framework for the development strategies of the state and regions and it provides the possibility of joint learning and the opportunity to abandon the illusion that the differences between the city and the country will be levelled automatically.

A very similar situation exists in the other developed countries of the world, e.g. rural America is experiencing revitalization and transformation. The United States Department of Agriculture notes that 'rural' is no longer synonymous with 'farms' and 'ranches'. Of the 60 million people living in rural America, 58 million are not farmers any more and, even among farm families, the majority of their income is obtained from non-farm sources. American rural communities have entered the age of economic diversification, modern transportation and instant communication. In 1994 the Department of Agriculture was reorganized and USDA Rural Development was created to administer non-farm financial programmes for rural housing, community facilities, water and waste disposal and rural businesses. At present, USDA Rural Development administers more than 40 programmes with investment in rural communities (USDA Rural Development, 2007).

Case Studies

1. Discuss in groups the main reasons for and results of rural development in your country (see Fig. 3.1).
2. What is the multiplier effect of agritourism? Try to find a few examples of the multiplier effect of agritourism development in any region of your country.
3. Are there any differences between multifunctional rural development and multifunctional agricultural development?

4 Agritourist Space

Definition

Agritourist activity takes place in a space, which we shall call agritourist space. In the narrow sense agritourist space refers to the area of a farm providing agritourist services, its natural landscape and the landscape of the expanse that is the result of human activity. It includes the architecture, the farm landscape and the scenery formed by production activity. It is also formed by the air and its quality. It is not only the owner that has influence on the agritourist space of the farm but also the near and more distant surroundings. Agritourist space in the broad sense refers to the near and more distant surroundings of the farm and it includes the landscape and architecture of the surroundings (e.g. architecture of the village) and the scenery resulting from the manufactured products, but it also refers to the cleanness of the air, water, smells, noise, etc. If the dominating element of the agritourist scenery is a village, we speak of village space; if it is agriculture, we speak of agricultural space.

Evaluation of Agritourist Space

Agritourist activity would like its space to be very attractive. Studies (Jolly and Reynolds, 2005) have shown that tourists assign value to individual elements of space. For inhabitants of big American cities, orchards are the most attractive, followed by forests, with grazing land ranking next, followed by fields. Farming buildings and facilities are considered the least attractive. Inhabitants of small towns have slightly different preferences. For them the forests are the most attractive, followed by orchards.

The agritourism valuation of an area refers to the estimation of its value in terms of its suitability for the development of tourism on the basis of qualitative and quantitative features of the natural and anthropogenic environment. It may best be done on the basis of traits referring to public statistics records. Some characteristics have an advantageous effect on the development of agritourism and these are called stimulants, others – nominants – are not related, and finally we have features with an adverse effect on the development of agritourism and they are called destimulants. Stimulants, for example, may include protected areas, forests, meadows and pastures, water bodies, orchards, roads, historical monuments, tourist infrastructure, diverse species of crops and farm animals. Destimulants are first of all connected with monotony, e.g. monoculture in plant cultivation, large animal farms, especially pig and poultry farms. The density of population together with the share of people earning a living from non-farming sources illustrates the intensity of urbanization and industrialization as well as pollution and noise coming from local sources. An example of a nominant may be moderate population density or moderate industry development.

The criteria for evaluation of agritourist space have been defined by some authors. This allows us to specify which space has better predispositions for tourism. Lane (1992) distinguishes six factors that determine the value of agritourist space. They are:

1. The value of the beauty of landscape, which includes: mountains, sea coasts, lakes, islands, picturesque river valleys and areas of specific value and specific beauty, such as swamp or forest.
2. Areas of wild nature and wilderness.
3. Cultural values, which include historical buildings, small towns, villages and places as well as ethnic heritage.
4. Particularly favourable conditions for hunting, fishing, skiing, hiking.
5. Good accessibility to a broad market of consumers.
6. Effective and professional promotional and commercial activity and proper management.

The first five factors can be characterized as objective values, inherent in an area and its properties. The sixth is the human factor. The marketing and management of tourist services may develop the housing and services base in a relatively short time. They may improve and extend it, though this requires more time and financial outlay. The natural and cultural values cannot be created again, though they can also be extended and improved. These values can be considered permanent factors of the quality of agritourist space, whereas the others are variable ones.

Some entire countries are predisposed to tourism and agritourism whereas in others, part of their territory is not particularly attractive. Methods of evaluation of agritourist space are needed. The method of evaluating agrotourism space developed by Drzewiecki (1992) includes mainly material elements, which are considered to be relatively stable. In this method criteria were selected so that they could statistically reflect important factors for rural tourism. A considerable role for the determination of agritourist regions based on statistical data is played by the method applying a taxonomic measure of development, called the Hellwig index (Wysocki and Lira, 2003). Figure 4.1 presents the division of the Wielkopolska region (Poland) into four classes of agritourist regions based on this method.

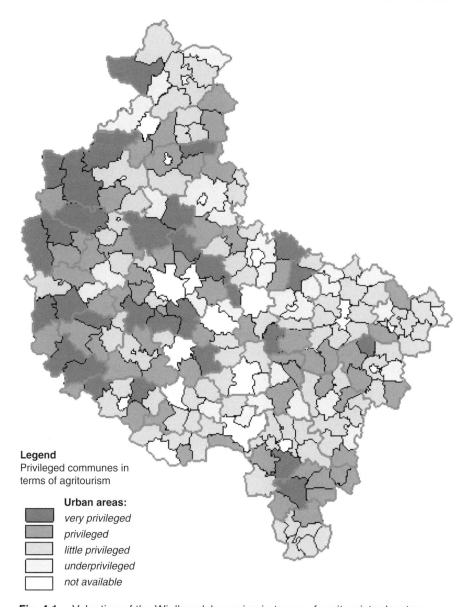

Legend
Privileged communes in
terms of agritourism

Urban areas:
very privileged
privileged
little privileged
underprivileged
not available

Fig. 4.1.   Valuation of the Wielkopolska region in terms of agritourist advantage.

Recognizing the value of agritourist space is very important for the development of agritourism. If somebody would like to establish an agritourist farm in an area that has no natural or architectural values, the number of people willing to participate in agritourism will be low. If a farm is located in an interesting agritourist space but it is devastated, for example, by the smell from a pig farm, it will not be attractive to agritourists either. Agritourist space is formed through centuries of human activity, but it is also affected by current human activity.

Fig. 4.2. The village of Estaing – from the list of the most beautiful villages in France; a beautiful stone bridge (photo by M. Sznajder).

Many countries draw up lists of the most beautiful places and areas. For example, in France there is a list of the most beautiful villages. It includes such villages as Saint-Côme d'Olt, Najac, La Couvertoirade, Sainte Eulailie de Cermon, Brousse le Château, Estaing, Belcastel, Sauveterre de Rouergue and Conques. Since 1992, the Most Beautiful French Villages Association has been operating, which decides which places will be included in the list of the most beautiful. There are similar lists in Italy, Greece, England and Ireland. A series 'The Most Beautiful Villages . . .' is issued. It is published by the Savvy Traveller from Chicago, Illinois (http://www.thesavvytraveller.com, accessed April 2008). Figure 4.2 shows a fragment of the village of Estaing, from the list of the most beautiful places in France.

The term 'tourism icon' also appears in the literature. A tourism icon is a particularly beautiful attraction that acts as a magnet attracting tourists to a given country. Every tourist should see the icon. For a particular country all methods of promotion of tourism induce the need to see the icon. In every country there are many beautiful places and making a decision concerning which of them deserve to be included in the list of the most beautiful ones is difficult. It also has specific consequences. Only a few countries have already decided to regard a beautiful place as a tourism icon (Becken, 2003).

The Types of Agritourist Space

The variability of agritourist space is enormous and its precise classification is a relatively complicated task. The variability of agritourist space depends upon its biophysical (i.e. including the climate, water relations, sculpture of terrain, geological structure, the animal world and vegetation) and anthropogenic qualities, which include:

1. Area configuration.
2. Forms of terrain.
3. Natural fauna and flora.
4. Type of land use.
5. Buildings.
6. Size of farms.
7. Specialization of production.

When characterizing agritourist space it is additionally necessary to specify its quality, which depends upon cultural and social values as well as the landscape and the quality of the air and water.

Area configuration

In respect of area configuration it is possible to distinguish the following principal forms of agritourist space:

1. Plain areas.
2. Undulating areas.
3. Hilly areas.
4. Mountainous areas.

Configuration determines the attractiveness of an area. Generally flat areas are considered less attractive than the other forms, as may be seen in the Canadian prairie, the Hungarian salt plain or the western Siberian lowlands. However, even flat areas have some charm. A certain book for schoolchildren in Canada describes the beauty of such flat areas. These lowlands extend in a wide belt from Paris through Belgium, Holland, northern Germany, Poland, Belarus, Russia to the Urals, only to be transformed further east in the vast east Siberian lowlands. Valleys of big rivers are also large flat areas. Sometimes plains are directly adjacent to mountains. Flat areas in South America neighbour directly on the high Andes, and in New Zealand such areas are adjacent to the Southern Alps. Mountainous areas also cover vast areas in Europe, Asia and the western part of both Americas.

Attractive relief forms

Naturally configured surface features such as mountains, valleys, gorges, waterfalls, rivers, lakes and warm springs create a specific background for agritourist space. New Zealand is rich in waterfalls. A calm narrow river flowing lazily across pastures turns into a powerful and dynamic waterfall after reaching a precipice. In spite of this abundance, waterfalls give variety to rural space only in a few places. More often they are hidden in mountain valleys. River valleys and lakes make a particularly interesting agritourist space. River valleys may form huge complexes of swamps. Swamp is very exciting: for example, the River Biebrza in Poland, whose valley forms a unique swamp complex, where human farming activity (cattle breeding) and wild nature exist side by side. Another type of landscape is formed by river canyons. Almost all rivers form their own gorges or river gaps. The most famous and biggest is the Grand Canyon. Rural areas may boast of small but picturesque gorges. In many countries of northern Europe and in Canada there is an abundance of lakes, which increases the value of agritourist space. The fauna and flora also affect the quality of space. People travel far to see some native species of plants and animals. In certain areas due to the unique character of the fauna and flora, national or scenic parks are established.

Type of land use

In respect of agricultural production five types of agritourist space can be distinguished:

1. Cropping space.
2. Livestock space.
3. Orchard and plantation space.
4. Forest space.
5. Fishing space.

Cropping space is formed by farms specializing in crop production. The aim of agricultural production can be commodity plant products or feed for animals kept in buildings. The first category in a moderate climate is dominated by cereals, potatoes, rape, sunflower, soybeans or sugarbeet, whereas the other category is dominated by fodder plants, mainly maize and lucerne, less frequently clover. In tropical countries cropping space is predominantly covered by other crops, such as rice, cotton, sugarcane, sorghum, etc. In Europe, due to the type of production, there is usually a mixed space, where many crops are grown near one another. The cropping space, conditioned by the types of plants grown, varies in time, i.e. the landscape constantly changes during the year. There is a difference between the winter landscape, the landscape of spring work and those of growing plants, blooming (rape looks particularly beautiful), ripening, harvest and autumn work (ploughing, sowing – the landscape before winter) – see Figs 4.3–4.6.

There are regions where, despite the fact that animal production prevails, animals cannot be seen in fields. This is the case of livestock farming based on feed coming from arable land. The feed harvested from fields is transported to buildings for livestock and therefore animals can be kept in them all year round. However, what appears in the landscape is characteristic buildings with structures for feed. In fields we can observe large expanses of maize grown for silage, which is left in the form of characteristic stubble for a longer period of the year. This type of landscape can be found for example in the areas of intensive livestock farming in Canada or the USA.

Fig. 4.3. A varying agritourist space depending on the seasons of the year: autumn – after harvest (photo by L. Przezbórska).

Fig. 4.4. A varying agritourist space depending on the seasons of the year: a field in winter (photo by M. Sznajder).

Fig. 4.5. A varying agritourist space depending on the seasons of the year: a field in spring (photo by M. Sznajder).

Fig. 4.6. A varying agritourist space depending on the seasons of the year: poppies blooming in early summer (photo by M. Sznajder).

The livestock farming type of agritourist space appears where farms are oriented towards animal production but animals are almost exclusively grazed rather than fed indoors. In countries of year-round vegetation, e.g. in Uruguay, Argentina, New Zealand and even western Europe, animals stay in pastures all the time. In the countries of northern Europe and North America, grazing of pastures is seasonal, i.e. in spring, summer and part of autumn. Meadows and pastures dominate the breeding type of agritourist landscape. Usually in such an area there is no arable land or only a small amount of it. The share of permanent grassland in land use in Europe is illustrated on the map in Fig. 4.7.

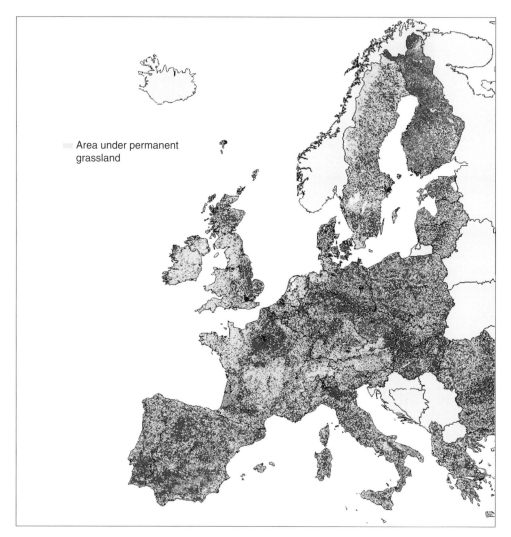

Area under permanent grassland

Fig. 4.7. Area under permanent grassland in utilized agricultural area (UAA) in the European Union (copyright European Environment Agency (EEA), http://www.eea.europa.eu, Copenhagen, 2001, in: http://dataservice.eea.europa.eu/atlas/viewdata/viewpub.asp?id=558 (accessed March 2008)). (reproduced with permission.)

The map shows areas of considerable percentage of permanent grassland in Europe. It pertains to the central and western part of the region, river valleys and especially western parts of France, the Massif Central and the United Kingdom and the whole of Ireland. In Argentina, New Zealand and southern Russia, large areas are almost exclusively dominated by pastures and the share of arable land is minimal. In European countries, the area of forests and permanent grassland has been decreasing in favour of arable land. An example of such a change in the years 1961–2005 in using land in France is shown in Fig. 4.8: the area of arable land is increasing steadily at the expense of permanent grassland.

Currently the situation is changing. In some countries the area of meadows, pastures and forests is increasing at the expense of the area of arable land. For example, in the USA (see Fig. 4.9) for a long time the area of permanent grassland and arable land was constant; however, in the last 15 years the area of permanent grassland has increased at the expense of arable land.

Pastures are organized in various ways. If they are private property, they are divided into larger or smaller sections. They can be separated from each other by wire fences, wooden fences, bushes or stones, which can often be observed in Ireland or England. Sometimes pastures are the property of a commune. Then the characteristic enclosures are missing and the cattle of different owners graze together. In pastures, various species of animals can be encountered: dairy cattle,

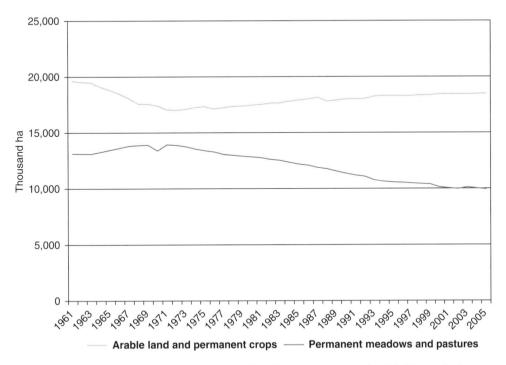

Fig. 4.8. Changes in the areas of arable land and permanent grassland in France in the years 1961–2005 (based on data from FAOSTAT 2008).

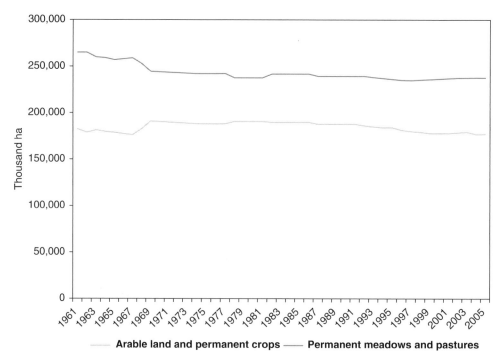

Fig. 4.9. Changes in the areas of arable land and permanent grassland in the USA in the years 1961–2005 (based on data from FAOSTAT 2008).

beef cattle, sheep, horses, deer, llamas and alpacas, and sometimes geese. Mountain pastures of steep gradient and low yield are treated as pastures for sheep, and pastures in plains are used for milk cattle. The variability of this space and landscape in time is usually smaller than the variability of landscape in agricultural space and this concerns the period of occurrence of young animals, haymaking, the production of hay silage, etc.

For examples of the use of livestock space in Italy and Australia, see Figs 4.10 and 4.11.

Orchard space and plantations can be found in regions where fruit and so-called specialist crops are grown. They vary depending on the type of crop grown (Figs 4.12 and 4.13). They are very characteristic where olives, oranges or grapes are grown. The landscape of orchard space constantly changes during the year. There is a difference in its beauty at the time of the development of leaves, blossoming, ripening, harvest, leaves falling and in winter.

Fishing space can be found in areas of developed fish breeding in ponds. The dominating element of the landscape is ponds separated from each other by dykes and small dams. The landscape does not change much during the year. In principle, the winter landscape can be distinguished, when ponds are empty, from the summer landscape, when ponds are filled with water. Fishing is an interesting and varied element.

Fig. 4.10. A photograph showing livestock space: buffaloes, Italy (photo by M. Sznajder).

Fig. 4.11. A photograph showing livestock space: kangaroos, New South Wales, Australia (photo by L. Przezbórska).

Fig. 4.12. Orchard space: orange orchard, New Zealand (photo by L. Przezbórska).

Fig. 4.13. Orchard space: apple orchard, Poland (photo by L. Przezbórska).

Fig. 4.14. Forest space seen from within the forest: rainforest in New Zealand (photo by M. Sznajder).

Forest space is connected with agritourism. Owners of large forested areas have their own tourist infrastructure, e.g. in Poland a guidebook to forest tourist centres has been published, which lists almost 5000 accommodation places located in the forest tourist space. However, many farms also have both fields and forests within their boundaries. In the summer, work is to be done in the fields, while the winter is the time to work in the forest. For this reason the landscape around a farm includes a forest in the background, which changes along with the geographical position and the seasons of the year. Forest may be viewed from outside as an element of the view in the surrounding landscape (preferably from a viewing point) or it may be observed from within. Psychologically these are two different perceptions. Figures 4.14 and 4.15 show different views from within a forest.

Spatial development

In respect of type of development, agritourist areas can be divided as follows:

1. Without development.
2. With compact development.
3. With dispersed development.

In areas of compact development, all farms and other major buildings are situated along the main road that runs through the town or village. In the case of dispersed development, farm buildings are located in the centre of an expanse of

Fig. 4.15. Forest space seen from within the forest: a beech forest of the moderate zone (photo by M. Sznajder).

farmland. Both types of location can be found in many countries. India used to be, and for the most part still is, one huge scattered village, extending from the Arabian Sea to the Bay of Bengal. The polder areas in the Netherlands are another example of dispersed location. Figures 4.16 and 4.17 show fragments of maps of rural areas with compact location of structures (Fig. 4.16) and dispersed location (Fig. 4.17).

Figures 4.18 and 4.19 show bird's-eye views of different locations: compact location of farms and dispersed one.

Size of farms

The size of farms also has an impact on the quality of agritourist space. The division of farms in terms of their size depends on the region of the world. What in one part of the world is considered to be microscopic in another is sufficient to sustain a family. In Europe the following types of space can be distinguished:

1. Very small farms – up to 5 ha.
2. Small farms – from 5 to 20 ha.
3. Medium farms – from 20 to 50 ha.
4. Big farms – from 50 to 100 ha.
5. Very big farms, whose area is bigger than 100 ha and sometimes even exceeds 2000 ha.

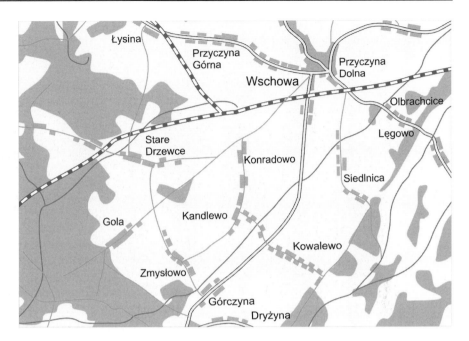

Fig. 4.16. A fragment of a map of rural areas with compact spatial development (redrawn from Atlas samochodowy Polski, 1999).

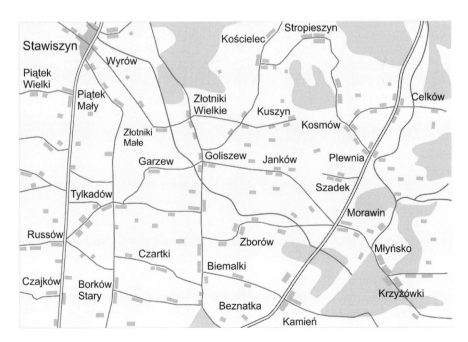

Fig. 4.17. A fragment of a map of rural areas with dispersed spatial development (redrawn from Atlas samochodowy Polski, 1999).

Fig. 4.18. Aerial photograph of rural areas: compact development of rural areas on Polish–Slovak border (photo by M. Sznajder).

Fig. 4.19. Aerial photograph of rural areas: location of a farm in the centre of an expanse: near Christchurch, New Zealand (photo by M. Sznajder).

The structure of farm size within a country is formed over a long historical process and is subject to constant changes. At some historical moments, the changes took place rapidly. The change of structure of farm sizes is conditioned by numerous factors, but above all by political and economic factors. At present, there is a process of polarization in the size of farms, i.e. the number of medium farms is decreasing and the number of big ones is increasing. The number of very small farms is relatively stable.

Mixed space, formed by farms of different areas, can also be distinguished. From the point of view of agritourist value, the smaller farms make the space more attractive. The beauty of such space is particularly visible in a bird's-eye view (Figs 4.18 and 4.19). Europe is an area dominated mostly by small and medium farms, though there are also areas where big farms are dominant, e.g. in the north of Germany, Poland and the former Soviet Union. In North and South America, Australia and New Zealand big farms prevail, whereas in Asia and Africa the farms are usually small units.

Specialization of production

In agricultural production monocultural space and diversified space can be distinguished. In monocultural space we observe only one landscape, for example endless fields of maize. In diversified space there are several types of cultivated plants in neighbouring fields. Usually diversified agritourist space is more interesting than monocultural production.

Monotonous monocultural agritourist space can be found in North and South America, where vast expanses are dominated by plantations of cereals, maize, soya and cotton. A similar situation is characteristic of Australia. In other parts of the world the space tends to be diversified.

The economic changes around the world that have taken place in the last two decades have caused agricultural space to become increasingly monotonous, which can be observed especially in those places where far-reaching specialization has occurred. Farm animals cannot be seen in pastures, as they are usually kept in buildings for livestock. The area of potato and beet plantations is decreasing, whereas the area of cereal plantations is increasing.

Agritourism and Protected Areas

Agritourism activities also develop within different kinds of protected areas in the world, that is, locations that receive protection because of their various environmental, cultural or other values. Most protected areas play a dual role, offering a refuge for wildlife and serving as popular tourist areas. Managing the potential for conflict between these two roles can become problematic, particularly as tourists often generate revenue for the areas, which, in turn, is spent on conservation projects. The largest protected areas in the world are in the United States, Greenland, Australia, Canada, China, Brazil and India. However, Greenland, Ecuador, Denmark, Venezuela, Germany, Austria and New Zealand have the largest share of protected areas in their total land territories.

The International Union for the Conservation of Nature and Natural Resources (IUCN, or the World Conservation Union), the largest professional global conservation network, defined a protected area as 'An area of land and/or

sea especially dedicated to the protection and maintenance of biological diversity, and of natural and associated cultural resources, and managed through legal
or other effective means.' IUCN categorizes protected areas by management
objective and identifies the following distinct categories:

- Category Ia – strict nature reserve;
- Category Ib – wilderness area;
- Category II – national park;
- Category III – natural monument;
- Category IV – habitat/species management area;
- Category V – protected landscape/seascape;
- Category VI – managed resources protected area;
- Category other – all protected areas that do not fit into one of the IUCN
  categories.

National parks are a protected area of IUCN category II. In 2004 there were
62,137 protected areas in the world, including 9568 nature reserves, wilderness
areas and national parks (Table 4.1). The total territory of all protected areas in
the world is more than 806.7 million hectares and is 6.1% of the total land area.
Globally, national programmes for the protection of representative ecosystems
have progressed. Some areas are considered particularly valuable with respect to
scenery or nature and therefore scenic and national parks are established there. A
national park is a reserve of land, usually, but not always, declared and owned by
a national government, protected from most human development and pollution.
Some countries also designate sites of special cultural, scientific or historical importance as national parks, or as special entities within their national park systems.
Other countries use a different scheme for historical site preservation. In many
countries, local government bodies are responsible for the maintenance of park
systems. Some of these are also called national parks. Figure 4.20 shows the growth
in nationally designated protected areas in the world between 1872 and 2006.

The largest national park in the world is the North-East Greenland National
Park, which was created in 1974 (972,000 square kilometres), and the world's first
national park, established as early as 1872, is Yellowstone National Park, USA.
There soon followed parks in other nations, e.g. in Australia (the Royal National
Park in 1879), in Canada (Banff National Park, then known as Rocky Mountain
National Park, in 1885), in New Zealand (Tongariro National Park in 1887), and
finally in Europe the first national parks were established in Sweden in 1909. The
first national park in Africa was designated by the government of South Africa in
1926 (Kruger National Park). After the Second World War, national parks were
founded all over the world. An up-to-date list of national parks in the world is available on the website National Parks Worldwide (accessed March 2008).

There are also different kinds of scenic and landscape parks in various countries in the world. They can be described as areas of outstanding natural beauty,
where the interaction of people and nature over time has produced an area of
distinct character with significant aesthetic, ecological and/or cultural value, and
often with high biological diversity. Almost every country designates national parks
and other categories of protected areas. Everywhere they create a very special
tourist area and attract numerous tourists. For example, it is estimated that in 2006

Table 4.1. Protected areas in the world, 2004. (From Earth Trends, 2004; World Resources, 2005.)

Regions	All areas under IUCN management categories I–V (2004)[a]			Nature reserves, wilderness areas and national parks categories Ia, Ib and II		Number of protected areas larger than 100,000 hectares
	Number of protected areas	Total area (1000 ha)	Percentage of total land area	Number	Area (1000 hectares)	
World	62,137	806,722	6.1	9,568	396,205	2,178
Asia (excl. Middle East)	3,999	191,450	7.9	729	44,239	387
Europe	40,318	137,694	6.1	2,329	36,725	337
Middle East and North Africa	503	33,360	2.7	81	15,237	–
Sub-Saharan Africa	913	142,025	5.9	299	78,542	425
North America	6,294	131,738	5.9	2,666	99,952	347
Central America and Caribbean	547	6,041	2.2	214	4,628	45
South America	1,195	106,018	5.9	557	66,424	466
Oceania	8,368	58,396	6.9	2,693	50,457	117
Developed countries	55,792	353,555	6.3	7,798	197,483	843
Developing countries	7,242	454,467	5.9	1,775	199,026	1,335

[a]Marine and littoral protected areas are excluded from these totals.

the USA national parks were visited by 273 million people. There are 391 national parks in the United States and they cover 84 million acres.

In comparison, there are 23 national parks in Poland with a total area of 3285.8 square km, which cover approximately 1% of the country's area. Polish national parks are exceptional in Europe for their range of wildlife, their size and varying geographical interest. Besides implementation of the main goal – nature conservation – the parks carry out scientific, education and tourist activities. The latter enable all interested people to experience direct contact with unique nature. This privilege, however, implies the obligation of conforming to every rule in the area of a national park. There are about 2000 km of nature trails in the national parks. Apart from the most common activity, i.e. hiking, the areas of national parks are also accessible for skiing, mountain climbing and canoeing. Educational and informational activity is carried out through preparation of so-called

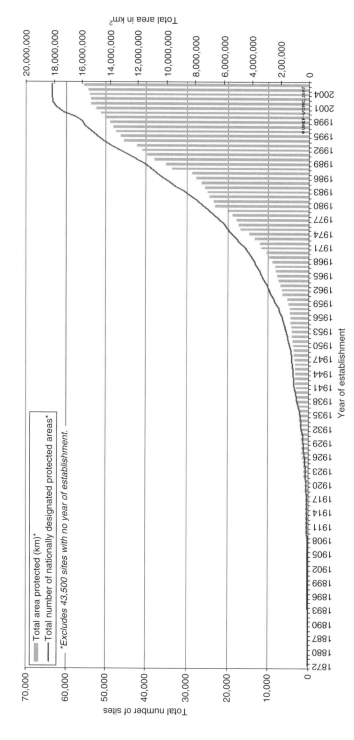

Fig. 4.20. Growth in nationally designated protected areas (1872–2006) (from World Database on Protected Areas (WDPA), World Conservation Monitoring Centre (UNEP – WCMC) and the IUCN World Commission on Protected Areas, 31 January 2007), in http://sea. unep-wcmc.org/wdbpa/PA_growth_chart_2007.gif (accessed March 2008).

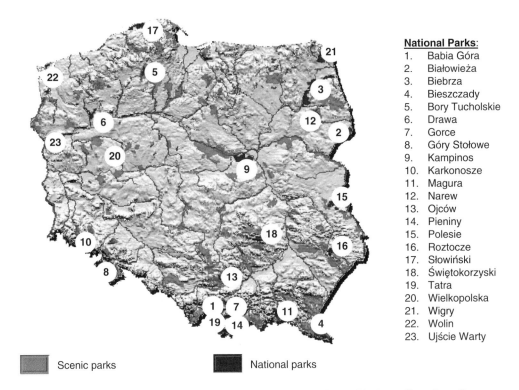

National Parks:
1. Babia Góra
2. Białowieża
3. Biebrza
4. Bieszczady
5. Bory Tucholskie
6. Drawa
7. Gorce
8. Góry Stołowe
9. Kampinos
10. Karkonosze
11. Magura
12. Narew
13. Ojców
14. Pieniny
15. Polesie
16. Roztocze
17. Słowiński
18. Świętokrzyski
19. Tatra
20. Wielkopolska
21. Wigry
22. Wolin
23. Ujście Warty

Scenic parks   National parks

Fig. 4.21. The distribution of national and scenic parks in Poland in 2006 (from http://www. mos.gov.pl/soe_pl/rys16b.htm, accessed April 2008).

'educational trails'. New forms of environmental education and related activities are being developed and tested there as well.

The total area of protected terrain in Poland in 2006 amounted to 100.4 thousand square km, i.e. 32.1% of the total area of the country. In addition to national parks, Poland also has landscape (scenic) parks, areas of protected landscape and natural reserves. Among these, landscape parks deserve special attention. These areas include examples of outstanding landscape, boasting aesthetic, historical and cultural values. By the end of 2006 in Poland 120 scenic parks had been established, occupying an area of over 24.5 thousand square km, which amounted to 7.8% of the country's area. The map in Fig. 4.21 shows the distribution of the national and scenic parks in Poland. In the vicinity of national parks agritourist activity is developing rapidly in support of the tourist process. More information on Polish national parks can be found on the Polish National Parks website (accessed March 2008).

Management of Agritourist Space

Managing a tourist-friendly agritourist space is extremely important for the development of agritourist activity. Developing agritourist space is not only aesthetic or economic but also a cultural issue. A correctly developed space encourages a visit to the region. Developing the space requires outlay from the budget of both the local community and individual farms. A lake with contaminated water or odour from a pig farm causes a considerable decrease in the quality of the agri-tourist space.

Certain types of business activity reduce the quality of agritourist space to such an extent that tourists are not willing to stay in the area. Farm animal production, especially large pig farms, lower the quality of the space to such an extent that agritourist activity can be excluded. An economic analysis made for a specific rural region should define which is more profitable for the relevant people: developing agritourism or, for example, animal production. If the economic analysis points to agritourism, animal production should be constrained. Where there is no chance for the development of agritourism, the idea of developing agritourist space does not have to come into the foreground.

Mahé and Ortalo-Magne (1999) presented a method of analysis to examine the competitiveness of agritourism in relation to intensive animal production. The authors consider the total economic effects of pork production or agritourism depending on the methods of the state's intervention in environment protection. It is a methodological contribution to the economics of managing agritourist space. The authors compared the economic effects resulting from the introduction of regulations concerning the environment protection for two regions – a pig-production region and an agritourist region. Finally they conclude that environmental regulations should have a local rather than a global character for all the space of the European Union. Local regulations correspond better to the needs of rural communities.

Zoning regulations as an instrument of development of rural areas

According to Mahé and Ortalo-Magne (1999), continuous limitations resulting from increased demands concerning the environment are an inappropriate constraint. They argue that intensive agricultural production should be allowed on the best soils. It is advisable to set local standards concerning the environment that better correspond to the local economic conditions. The restrictions concerning pig production in the pig-production region beyond the necessary health and safety standards adversely affect the local farms. On the other hand, the farms in the agritourist region profit from stricter standards. Despite the fact that each farm can observe that the standards lower the individual profit, overall each company benefits from it because nobody is allowed to destroy the local environment and its tourist attractiveness.

Entertainment and tourism are the main income potentials from green zones. Green zone – a term applied by Mahé and Ortalo-Magne (1999) – refers to the areas privileged to develop agritourism and those necessary for special environmental protection. They correspond to the protection areas and the national parks as well as those areas that are suitable for providing agritourist services. By introducing strict environmental standards, the agritourist potential is not endangered by the negative effects of intensification of agriculture or industry. Green zones enable protection of the areas of particular natural beauty and develop the recreational potential of various other forms of rural tourism. Green zones can act as protective areas for fauna and flora. If the size of these zones exceeds the area necessary for balanced food production, then they can be reduced.

Figure 4.22 shows a combined production effect of pig and agritourist farms in a region with comparative dominance of pigs in a situation with no environmental limitations (continuous line) and in a situation with environmental limitations (broken line). Figure 4.23 is a similar analysis in a region with comparative dominance of agritourism. Legal regulations will cause a drop in the pig production to level H. Taking into consideration the prices in the first case, regulation is not favourable, whereas in the other case it is favourable for both types of farms.

Each anthropogenic activity affects the development of agritourist space. Certain solutions favour increasing the space quality, while others degrade it. Buildings and their architecture, roads and structures affect the quality of agritourist space for a long period of time. Buildings may increase but they may also degrade the quality of agritourist space. The landscape of the space occupied by former state farms, where in the 1970s and 1980s apartment blocks were built, is miserable. The blocks considerably degraded the agritourist space. Similarly the agritourist space of many regions of Poland was degraded by unattractive-looking one-family houses resembling boxes. Houses with flat roofs are not a very interesting element of the landscape. Rural architecture is very significant. A building affects the agritourist space for many hundreds of years. Therefore there

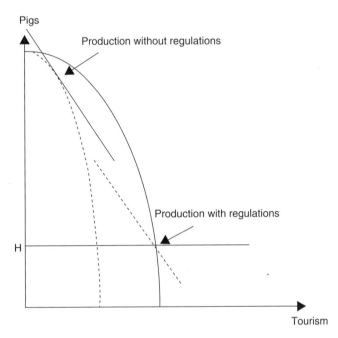

Fig. 4.22. An analysis of the competitiveness of agritourism and intensive pig production. Region 1, comparative advantage of pig production. (Compiled on the basis of Mahé and Ortalo-Magne, 1999.)

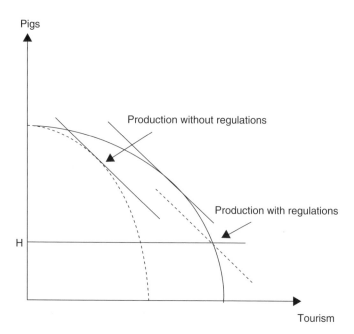

Fig. 4.23. An analysis of the competitiveness of agritourism and intensive pig production. Region 2, comparative advantage of agritourism. (Compiled on the basis of Mahé and Ortalo-Magne, 1999.)

is a good argument to regulate the construction of buildings that could potentially degrade the landscape. The landscape is disturbed by all types of poles and areas for industrial development, though the latest tendencies also favour matching industrial development with the landscape. Technological solutions configure the space for a shorter period of time, e.g. orchards configure the space for about 10–20 years. In southern France the configuring of landscapes is of such importance that the techniques of cutting grapes are regulated. Different techniques of cutting grapes result in different landscape effects.

Certain actions configure the agritourist space for a short period of time: a few days or even a few hours. The landscape changes especially at the time of blooming or harvest. Due to the change of production technology the landscape after harvest also changes. As recently as 20 years ago there used to be a beautiful post-harvest landscape. Sheaves of cereals were arranged in lines of 15 and a wagon transported the sheaves to the barn. Today what we see is usually straw in the form of pressed cubes or large rolled bales lying on the stubble.

It is not only the economic aspect but also the need for aesthetics that speaks in favour of good organization of rural space. The inhabitants of areas of particular agritourist values often realize that by taking care of the space they create for themselves better conditions for development in all respects. Not only is the development of agritourist space and rural space significant for the development of agritourism but it is also a sign of the overall culture of local inhabitants. The quality of agritourist space is a result of human activity. The specific condition of

agritourist space is a combination of the activity of all business entities and people functioning in a given area, especially farms and the companies of the agricultural and food industry.

Agritourist Space in Europe

Agritourist space in Europe is highly diverse in terms of surface features, flora
and fauna, type of architecture, farm size, the type and character of plant and
animal production, as well as variation in the length of night and day, depending
on the season of the year, and, what is equally important, considerable climatic
differences. This results in Europe being a very attractive area for agritourism.
There are vast lowlands, rolling plains and mountainous areas, as may be seen
by just looking at a map of this continent. The European Environment Agency
(EEA) developed a map (Fig. 4.24), presenting dominant types of landscape.
Obviously, landscape is only one of the elements of agritourist space. EEA dif-
ferentiated seven types of landscape, i.e. urban dense areas, dispersed urban
areas, broad-pattern intensive agriculture, rural mosaic and pasture landscape,
forested landscape, open semi-natural or natural landscape and composite landscape. Analysis of this map showing dominant types of landscape confirms the
common notion that the space of Europe is like a mosaic, i.e. highly diverse.
Neighbouring areas may be drastically different in terms of landscape. At times
there are also vast complexes of uniform landscape, e.g. pastures in the north-
east in Latvia, Lithuania or Estonia, in the west in Ireland, central France and the
Balkans – Croatia, Bosnia and Herzegovina. There are considerable forest complexes and areas under intensive agricultural use, in which large areas of natural
landscape were retained.

Moreover, natural flora and fauna change when we move from the north to
the south and from the west to the east. This pertains also to cultivated crops and
kept animals. For instance, when travelling from the north to the south, species
of farm animals change, starting with reindeer in the north, through dairy cattle
and pigs kept in central and southern Europe, to goats, donkeys or even buffaloes (Italy, Romania, Bulgaria) in southern Europe. Sheep and fat stock are kept
in the belt running from north-west (Ireland, UK, France) to the south-east
(Greece, Balkans, Italy). Goats predominate in the south. If to this variation we
also add variation in land use, traditional architecture, culture, religion and languages, then Europe must be seen as extremely rich in diverse space, which
determines its significant attractiveness for agritourism.

An Example of Changes of Agritourist Space

Agritourist space is subject to continuous changes. The changes are induced by
human business activity. Buildings are particularly significant as they form the
landscape for hundreds of years. The choice of direction of production changes
the landscape for many years. Agritourist space, especially the landscape,
changes depending on the season of the year (winter and summer landscapes),

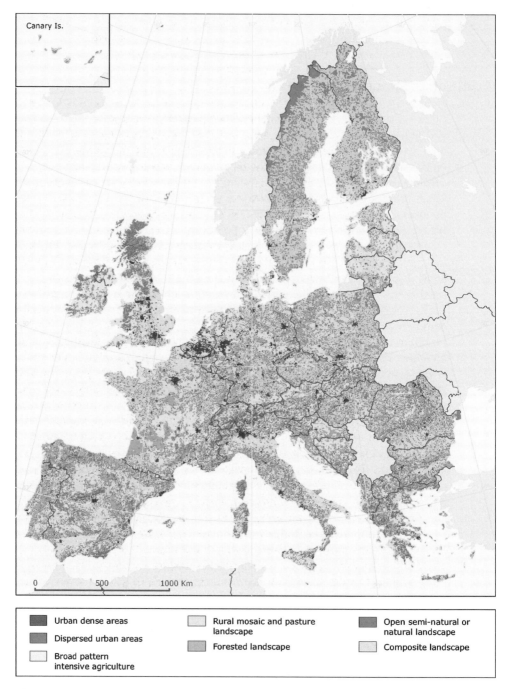

Canary Is.

0   500   1000 Km

	Urban dense areas		Rural mosaic and pasture landscape		Open semi-natural or natural landscape
	Dispersed urban areas		Forested landscape		Composite landscape
	Broad pattern intensive agriculture				

Fig. 4.24. The dominant landscape types of Europe (reproduced with permission) (copyright European Environment Agency (EEA), http://www.eea.europa.eu, Copenhagen, 2005, in: http://dataservice.eea.europa.eu/atlas/viewdata/viewpub.asp?id=2573 (accessed March 2008)).

the physiological phase of plant development (the blooming and ripening of plants), the time of the day (driving the cattle into a pasture in the morning, etc.) and the type of field work (haymaking, ploughing).

The last 50 years have seen numerous changes in the manner and technologies of production and land use. They have affected the natural structure and the functioning of ecosystems composing agritourist space. Łysko (2002) analysed the changes of this space in the area situated in the Drawsko and Wałcz Lake District. When discussing the genesis of the agricultural area he states that the original landscape of Poland had a typical forest character. Along with the development of agriculture the landscape began to be transformed into an agricultural landscape, resulting in deforestation of a large area of the country. At the turn of the 10th century the area of crops occupied only 20% of the area of contemporary Poland. By the 14th century their share had increased to 35%. By the end of the 18th century the forest area of Poland had decreased by more than 50%, whereas the area of arable land had increased. Further increases in the share of arable land intensified in the 19th century and continued until the mid-20th century. In 1946 the total area of arable land occupied as much as 66% of the country's area. In the last 50 years the area of arable land has decreased by about 3 million ha and in 2002 it was 54% of the total area of the country. These changes were the consequence of the afforestation completed after the Second World War, especially in West Pomerania.

After the Second World War, there were changes in the landscape resulting from the transformation of private farm fields into state farms, which were established in the former communist bloc in the estates taken away from landowners (for 40 years they were ineffective forms of management; in the 1990s they were mostly privatized or leased). They consisted in the transformation of the single cornfield arrangement of fields into large cultivated areas of more than 10 hectares. The increased industrialization and concentration of agricultural production, especially animal production, led to a range of unfavourable changes in the farming landscape. In a large part of the Drawsko and Wałcz Lake District significant ecological values were eliminated, e.g. trees in the middle of fields were cut down and minor water ponds were buried. What also contributed to the disappearance of ecological values was the decision to treat field forests and marshes as areas that are home to harmful plant and animal species. More recently it has been found that there is little correlation between field weeds, diseases or animal pests and ecological arable lands.

In the last 50 years the changes in the agricultural landscape have mostly resulted from forestation of drainage works. A consequence of forestation was that the species structure of forests underwent a considerable change. The share of coniferous species has grown considerably. Since the beginning of the 1990s the main reason for the reduction in the area of arable land is their idleness. The dissolution of state farms resulted in the idling of large areas of land, where secondary succession took place (ecological succession occurs where the biota has been destroyed but the soil has been preserved). It is certain that the agricultural landscape of this area will be subject to further changes.

The decrease in the area of arable land was also caused by its transformation into grassland. In the area of the end moraine near Czaplinek there have

been crucial changes in the spatial and quantitative distribution of potential ecological land appearing in the agricultural landscape. The number and area of
field forests have changed the most. Their area has increased by nearly 50%.
The number of small forests, with an area up to 0.1 ha, has changed a lot. Most
emerged in the direct vicinity of arable lands and idle lands, near drainage ditches
or in grassland and near ponds. The density of field forests per square kilometre
also increased from 7 trees per sq km in 1948 to 9 trees in 1997.

The analyses of changes in the number of ponds showed that, of the 397
ponds listed in 1948, only 179 remained by 1997. The main reason for the disappearance of ponds was lowering of the water level or eutrophication (a complicated biological process connected with water reservoirs that are lakes). As a
consequence of direct human activity, levelling the area of arable land eliminated
numerous ponds. As a result of forestation work, a large number of ponds became
located among forest ecosystems. Simultaneously in this area 150 new ponds
emerged, most of which occurred in wet meadows, usually as a result of plants
growing over the draining systems. The new items were small and usually their
area did not exceed 0.3 ha.

In this period the area of developed terrain increased by about 40%. In the
1960s and 1970s around Czaplinek several new state farm buildings were constructed. The buildings were raised mainly in the areas that had been arable
lands, which contributed considerably to the changes in the agricultural landscape and ecosystems in the examined area. In the last 50 years no crucial
changes in the development of individual villages have taken place, whereas a
large number of solitary buildings scattered around fields disappeared from the
agricultural landscape. The analysis of Łysko unarguably points to the enormous role of the human and economic systems in the development of agritourist
landscapes.

Case Studies

1. Based on your knowledge of the region, prepare the agritourism characteristics of space in that region, taking into consideration surface features, flora and
fauna, the type of architecture, farm size, cultivated plants and kept animals.
2. Decide what advantages and what difficulties result from the location of an
agritourist farm in protected areas.
3. Based on information collected from elderly people, photographic and press
documentation, follow changes in the agritourist space in your region in the period of at least the last 25 years.
4. Based on spatial development plans found in offices of state administrative
bodies and local government bodies, try to predict changes to be expected in the
agritourist space in the near future.

II   The Economics and Management of Agritourism

5 Agritourism Ventures

All business activities involve business ventures or enterprises. Business enterprises working in agritourism will be referred to as agritourism entities/agritourism ventures. Agritourism entities participate in the agritourism process, which comprises planning, production, marketing, sales and consumption of agritourism products and services.

Figure 5.1 shows the division of ventures in respect of their relationship with agritourism. The entities functioning in agritourism can be divided into four groups. The first group comprises the ventures directly providing agritourism products and services. Most frequently these are agritourism farms. The second group consists of the entities that use agritourism as a channel for direct sales of their products. Both farms and small agricultural and food-processing plants belong to this group. The third group consists of the ventures that use agritourism to promote their products. It applies to agricultural and food industry enterprises that enable observation of the production process or the organization of special places where they promote their products. The fourth group comprises the entities supporting (facilitating) the agritourism process. Such entities manufacture products for agritourism, e.g. souvenirs, folders, food. The group also comprises the entities that facilitate attracting tourists, such as travel agencies, and those facilitating the tourist traffic, e.g. transport companies.

Along with farms, agritourism ventures work towards their fundamental goal of making a profit in accordance with economic principles. They can adopt different legal forms, which also include non-profit organizations (NPO), which aim at facilitating the agritourism process or the marketing process in agritourism. There are many legal forms that enable agritourism ventures to run a business:

- Business activity on one's own account;
- Different kinds of registered partnerships, including general partnerships, limited partnerships and limited liability partnerships;
- Cooperative societies;
- Associations.

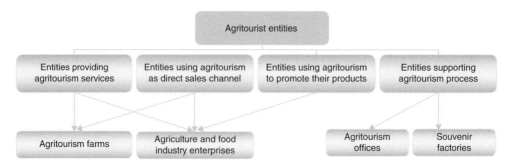

Fig. 5.1. A division of entities in agritourism.

Agricultural farms or ranches are a principal group of agritourism entities around the world. Besides plant and animal production they also provide agritourism products and services. The concept of agritourism was created especially for farms in order to increase their profitability.

In recent years agritourism has become relatively attractive to other legal forms of business activity. Households situated in rural areas that are not farms but have the necessary resources to provide agritourism services are also involved in agritourism. They may have various legal forms, but usually function as businesses run on one's own account. In the case of large-scale providers of agritourism products and services more advanced legal forms are established as registered partnerships, limited partnerships, limited liability companies, public limited companies and sometimes also cooperative societies. An agritourism household can be included in the structure of these legal forms.

In most countries of the world, a general partnership, or simply a partnership, refers to an association of persons or an unincorporated company. A general partnership is a company of people running an enterprise as their own company. The partnership can be formed by two or more persons, the owners are all personally liable for any legal actions and debts the company may face, and it is created by agreement and proof of existence. The assets of the partnership are owned on behalf of the other partners, and they are each personally liable, jointly and severally, for business debts and taxes. Profits or losses are usually shared equally amongst the partners. Each partner is liable for the company's liabilities without limit and with all his assets in solidarity with the other partners. Therefore the partnership can be defined as a type of business entity in which partners (owners) share with each other the profits or losses of the business undertaking in which all have invested. The company's assets are all the contributed shares or company purchases made during its existence. Each partner is entitled to represent the company. A partner's entitlement to represent the company applies to all court and non-court procedures of the company. However, depending on the partnership structure and the jurisdiction of the country in which it operates, owners of a partnership may be exposed to greater personal liability than they would as shareholders of a corporation. There has been considerable debate in many countries as to whether a partnership should remain an aggregate or be allowed to become a business entity with a separate legal personality. For example, France, Luxembourg, Norway, the Czech Republic and Sweden grant some degree of legal personality to commercial partnerships. Other countries, such as Belgium, Germany, Switzerland and Poland, do not allow partnerships to acquire a separate legal personality, but permit them the right to sue and be sued, to hold property and to postpone a creditor's lawsuit against the partners until he or she has exhausted all remedies against the partnership assets. In England and Wales a partnership does not have a separate legal personality; however, in Scotland partnerships do have some degree of legal personality.

A limited partnership is a company whose goal is to run an enterprise as their own company, where at least one partner (general partner) has unlimited liability to creditors and the other partner has limited liability (limited partner). The general partners are, in all major respects, in the same legal position as partners in a conventional venture, e.g. they have management control, share the right to use partnership property, share the profits of the firm in predefined proportions and have joint and several liability for the debts of the partnership. The limited partners have limited liability, which means that they are only liable for debts incurred by the firm to the extent of their registered investment, and they have no management authority. The earliest form of limited partnership arose in Rome in the 3rd century BC. The Napoleonic Code of 1807 reinforced the limited partnership concept in European law. In the United States, limited partnerships became widely available in the early 1800s, although a number of legal restrictions at the time made them unpopular for business ventures. Limited partnerships are distinct from limited liability partnerships, in which all partners have limited liability.

A limited liability partnership, abbreviated to ltd, can be established for business purposes by one person or more. A limited liability partnership comprises elements of partnerships and corporations. Limited liability results from the fact that partners' individual property is not liable for the company's liabilities. When a limited liability company is established by only one person (a sole partner), the person executes all rights that an assembly of partners is entitled to. Usually a prerequisite for establishing a limited liability company is collecting the company's capital to a certain amount and entering the company in a register. A limited liability company is managed by the assembly of partners, a management and supervisory board and an auditing committee.

A public limited company is established on condition that the partners collect the share capital to a certain amount covered in cash or in-kind contributions and divided into shares of equal nominal value. There is a minimum share capital for public limited companies, of which at least a quarter must have been paid for. For example, in the UK before a public limited company can start business, it must have allotted shares to the value of at least £50,000 (in the Republic of Ireland €38,092.14 and in Poland 50,000 zlotys). In return for the contributed capital, the partners receive shares, which are securities identifying their owner as a member of the company, or a shareholder. Each share can have a registered character or be issued to bearer, and it is transferable. In addition, shares can be divided into ordinary and preference shares. The latter give special rights, e.g. to vote, receive dividend or divide the stock if a public limited company is liquidated.

Establishing a public limited company requires that the founders should sign the statute. The company becomes a legal entity when it is entered in a special register and from that moment on as a legal entity it is liable with all its stock. Shareholders are not individually liable for the company's liabilities. The authorities of a public limited company are usually a general assembly, the management and the supervisory board. In general, a public limited company is a type of limited company whose shares may be offered for sale to the public.

The position of the cooperative sector in agritourism is of little importance, though cooperative traditions in agritourism are very rich. Cooperative movements originated mainly at the end of the 19th and in the first half of the 20th century. As historical data show, the cooperative movement originates in poor societies. Newly established cooperative societies are usually small. The member-founders have a full sense of egalitarianism. Nowadays cooperative societies are very rarely established, as other forms of ownership are more attractive to investors. Agritourism is developing in the period of economic history when the importance of the cooperative movement is declining. Mass establishment of agritourism cooperative societies is not very likely to occur in the 21st century.

A cooperative society (often referred to as a co-op) is a voluntary association of an unlimited number of people of alterable membership and with an alterable share fund, who run a joint business activity in the interest of their members. A cooperative may also be defined as a business owned and controlled equally by the people who use its services or who work in it. The prerequisite of establishing a cooperative society is to pass and sign a statute by the member-founders, which must be in agreement with the regulations of the cooperative law. A cooperative society must have a certain minimum number of members. The organs of a cooperative society are the general assembly, a supervisory board, the management and assemblies of member groups. The supreme organ in a cooperative society is the general assembly of members or, in the case of too large a number of members, the assembly of delegates. A general assembly is usually held once a year. It elects the organs of the society – the supervisory board and auditing committee – and votes on the most important resolutions. The supervisory board is appointed to control and supervise the society's activity. It is elected by the members for a certain number of years. The board is headed by the president. The supervisory board appoints the society's management, which is headed by the management chairman. The management manages the company's affairs without delay and represents it outside. The general rules of the cooperative movement are one member, one vote, and the cooperative society is an organization open to new members. The society's assets are its members' private property. As a legal entity, during its existence the cooperative society is the exclusive owner of its assets; however, if it is dissolved, only the members are entitled to the remaining assets. Agricultural cooperatives are widespread in rural areas around the world. For example, in the USA there are both marketing and supply cooperatives. Some of them are government-sponsored and they promote and may actually distribute specific commodities. In some European countries, there are strong agricultural and agribusiness cooperatives, as well as agricultural cooperative banks.

Table 5.1. The most important aspects of business entities that may function in agritourism.

Name of entity	Legal entity	Manner of establishment	Managing organs	Range of liability
Business activity on one's own account	None	Entering the register of enterprises	Owner, according to one's own decision	Full, all owned property liable
General partnership	None	Established as a result of a written agreement; must be entered in the register of enterprises	Each partner is entitled to represent the company	Each partner is liable for the company's liabilities without limit with all his assets in solidarity with the other partners
Limited partnership	None	Established as a result of a written agreement; must be entered in the register of enterprises	Managing the company is the exclusive competence of general partners whereas limited partners remain passive	General partners are liable to creditors without limit with all their assets, whereas limited partners are liable to the amount of limited partnership specified in the agreement
Limited liability partnership (LLP)	Applicable	Established as a result of an agreement made by partners, contributing shares to cover the company's capital, establishing the company's organs and entering the register of enterprises	Assembly of partners, supervisory board (or optionally auditing committee), management	The company is liable to creditors with all its assets
Public limited company (PLC)	Applicable	The company is established on condition that partners make an agreement, contribute shares to cover the company's capital, establish the management and supervisory board and enter the company into the register of enterprises	General assembly, supervisory board, management	The company is liable with all its assets, shareholders are not liable for the company's liabilities
Cooperative society (Co-op)	Applicable	Member-founders pass and sign the statute, elect the organs and enter the society in the register of enterprises	General assembly, supervisory board, management, meetings of member groups	Members are liable for the society's liabilities only up to the amount of the declared shares
Association	None	Statute, decisions of the management	General assembly, management, auditing committee	Association, all assets liable

An association is another legal form available to agritourism. The prerequisite to establishing an association is the declaration made by a group of people who intend to join the association. The association is a group of individuals who voluntarily enter into an agreement to accomplish a purpose. The association works in accordance with the rules specified in the statute. The statute is subject to registration in court. The managing organ is the general assembly of members. The association's assets can be membership fees, donations, legacies, bequests, income gained from its own activity and income from the association's assets and from public generosity. An association may receive subsidies. It may also run a business activity if its statute allows it. The income gained from the association's business activity serves the realization of the goals specified in the statute and it cannot be distributed among the members. A professional association or professional society is usually a non-profit organization that exists to further a particular profession, to protect both the public interest and the interests of professionals, e.g. an agritourism association.

Table 5.1 compares the most significant legal aspects concerning the form of property, manner of establishment, managing organs and range of property liability of business entities that can function in agritourism.

Case Studies

1. Check the types of organization that are allowed by law in your country.
2. List which forms of economic activity may best be applied in agritourism.

6 Regulations of Agritourism Activities

Starting a new business in any kind of agritourism venture one must be aware of the many rules and regulations that it is subject to. Like all business activities, agritourism activity is subject to a range of legal regulations, which must be abided by. Agritourism may be affected by a wide variety of different regulations. Various taxes, licences, codes and fees have to be carefully examined, and in some cases even permits are required. It depends on the location of the enterprise (country or even the region), attractions offered, whether employees are hired and a number of other factors (Agritourism in Focus, 2005).

Agritourism business activity can be affected by general business taxation, including personal income tax, business income tax, sales tax, property tax and business licence fees, as well as employee regulations comprising payroll tax, wage and hour regulations, hiring regulations, safety and health legislation and others. Business activity sometimes needs to be registered, and account settlement with the tax office is required. Prior to the registration of a business activity, receiving a business statistical number and tax identification number is necessary. Then opening a bank account, purchasing a stamp, registration at the Social Insurance Institution, etc., may be required. Companies functioning in the agritourism market may also be regulated by the Trading Companies Code.

Detailed legal regulations can be found in official journals and other legal regulations. The most significant legal problems related to agritourism activity are:

- The duties related to starting and registering a business activity;
- The duty to settle the income tax gained on the income from tourism and the commodities and services tax, i.e. value-added tax (VAT, levied on the added value that results from exchanges; an indirect tax, in that the tax is collected from someone other than the person who actually bears the cost of the tax), or goods and services tax (GST, i.e. a value-added tax imposed by several countries, including Australia, Canada, Hong Kong, New Zealand and Singapore);
- The sanitary duties related to renting rooms and catering;

- Fire protection duties;
- Environment protection, usually related to starting campsites and camping fields beyond a built-up area;
- Registration, statistical and other duties.

In general all regulations can be divided into three groups:

- General regulations, which are likely to have an impact on most agritourism enterprises (e.g. zoning regulation, i.e. land-use regulation, business licence and taxes, sales tax collection and remittance, etc.);
- Employment regulations, which have an impact on enterprises employing staff (fair labour standards, child labour regulations, occupational safety and health regulations, unemployment taxes, income tax, etc.);
- Permits and licences for specific types of enterprises, which have an impact on certain types of entities or types of attractions (food service permit and inspections, retail food store permits and inspections, winery licensing, fee-fishing regulations, etc.) (Agritourism in Focus, 2005).

As the regulations, permits and licences affecting agritourism vary around the world and may often be imposed at state and regional levels of government, only a limited number of examples of such laws and regulations will be discussed. However, one must remember that it is very important to investigate all regulations in order to avoid penalties, fines or interruptions to agritourism business (Agritourism in Focus, 2005).

One of the regulations that is likely to have an impact on agritourism enterprises is a zoning ordinance. Zoning is a system of land-use regulation in various parts of the world, including North America, the United Kingdom and Australia. The term 'zoning' derives from the practice of designating permitted uses of land based on mapped zones that separate one set of land uses from another. Zoning regulations affect the land use in a specific area and zoning may be regulated by cities or by counties. For example, in the USA special laws and regulations restricting the places where particular businesses should be carried out have been in force for a long time. They were developed at the turn of the 19th and 20th century, and by the late 1920s most of the nation had developed a set of zoning regulations that met the needs of the locality. However, some cities and counties do not have a zoning ordinance, e.g. Houston, Texas (New York City Zoning, 2008). In Tennessee, USA, a survey of 210 agritourism entrepreneurs, conducted in 2003, found that 27 (i.e. almost 13%) operators had difficulties with zoning for their enterprises (Bruch and Holland, 2004). Zoning ordinances may regulate the location, height and size of buildings and other structures, the percentage of a lot that may be occupied, the size of yards, courts and other open spaces, the density and distribution of population, the uses of buildings and structures, e.g. for recreation, the use of land for different purposes, including residence, recreation and agriculture, regulations for parking, etc. (Agritourism in Focus, 2005). Some areas of the USA are amending agricultural zoning regulations to permit some value-added agricultural products and activities, including agritourism (Retzlaff, 2004). In the United Kingdom there is a system of town and country planning, which comprises development control or planning control, through which local government regulates land use and new building.

Business registration is required by jurisdictions in most countries. A business licence can be a business registration, but many jurisdictions require further licences beyond registration. The business licence is a permit issued by government agencies that allows an individual or a company to conduct business within a certain area – the government's jurisdiction. Sometimes multiple licences are required, which are issued by multiple government departments and agencies, e.g. if an agritourism operation is located within city limits, the owner may need to obtain both a county business licence and a city business licence. The business activity and physical location (address) determine most licence requirements, as they vary between countries, states and local municipalities. Other determining factors may include the number of employees and the form of business ownership, such as a business activity on one's own account or a corporation. There are often many licences, registrations and certifications required to conduct a business in a single location (Agritourism in Focus, 2005). Sometimes registering a business activity involves the obligation to have social insurance at a social insurance institution and to pay premiums, which include old-age pension insurance, work incapacity insurance, illness insurance, accident insurance, a job fund and health insurance. Registering a business activity may involve changes in calculating local taxes, e.g. in Poland (Raciborski *et al.*, 2005).

All persons, corporations and other legal entities running a business are obliged to pay income tax, i.e. a tax levied on their financial income. There are various income tax systems around the world, with varying degrees of tax incidence (progressive, proportional or regressive). Usually businesses with annual sales of less than a certain value may be exempt from paying business tax (Agritourism in Focus, 2005). An income tax levied on the income of companies is often called a corporate tax, corporate income tax or profit tax. Individual income taxes often tax the total income of the individual, while corporate income taxes often tax net income. In Poland the specific element in legal regulations that individual farmers find particularly interesting is the regulation indicating when a farmer running an agritourism activity has a duty to clear accounts with the tax office as a person running a business on their own account. Complete tax exemption applies to the income gained on renting out guest rooms by persons who simultaneously meet the following conditions (Income Tax Act of 26 July 1991, Art. 21):

- Rooms are rented out to holidaymakers;
- The rented rooms are part of apartment buildings;
- The apartment buildings are situated in rural areas;
- The landlord has a farm and the buildings in which rooms are rented belong to the farm;
- The number of rented rooms is not higher than five.

The exemption also applies to the income gained on catering for the guests living in the rented rooms. Anyone who meets the conditions is entitled to the exemption. If somebody rents out more than five rooms, they cannot regard the income gained on the first five as exempt from tax and the others as taxed. They are under an obligation to tax all the rooms. In Poland a person renting out up to five rooms is obliged to register income for tax on commodities and services (VAT).

Besides legal entities, organizational units that are not legal entities and persons selling commodities or providing paid services in Poland are obliged to pay a tax on commodities and services. Value-added tax (VAT), or goods and services tax (GST), is a tax on exchanges. It is levied on the added value that results from each exchange. In a VAT system all businesses remit taxes on their sales but they are also refunded the amount of VAT remitted by their suppliers. Value-added tax differs from a sales tax because a sales tax is levied on the total value of the exchange. Sales and use taxes are imposed on the retail sale, lease or rental of tangible personal property, the gross charge for specific taxable services and the gross sales for amusements. The tax is imposed on the purchaser, but the seller is liable for collecting and remitting the taxes. Sales and use taxes are usually imposed by the state (Agritourism in Focus, 2005).

If the owner of an agritourism enterprise employs personnel for the activity, he is affected by several more regulations. They vary according to the type and size of the enterprise, the nature of the employee's job and the number of employed persons. An agritourism entity may be affected by regulations on the minimum wage, overtime compensation, equal pay, child labour and other ordinances applying to full- and part-time workers. There may also be some compulsory workers' compensation insurance for the employees and/or unemployment insurance taxes.

In many countries special permits or licences may be required for agritourism operation, depending on the type of activities being conducted. There are several permits and licences that are required for only specific types of attractions, e.g. (Agritourism in Focus, 2005):

- Food services permit and inspections (a permit for food service establishments, bed and breakfast, organized camps and other establishments serving food);
- Retail food store permit and inspections (a permit for a retail food store);
- Winery (a permit to produce, sell and transport wine) or liquor licence (a permit to sell alcoholic beverages);
- Petting zoos licence (an animal exhibitor licence);
- Nursery licence (a permit to grow, keep and propagate for sale and distribution nursery stock);
- Fee-fishing regulations (special regulations for fish farming and fee-fishing or catch-out operations), etc.

The regulations vary in individual countries and even counties. However, sometimes when running an agritourism business it is possible to avoid licensing, e.g. the New Zealand Association of Farm and Home Hosts (@home New Zealand) negotiated for several years with councils, government and different organizations the possibility of serving pre-dinner drinks of wine with a meal without a licence and the Association was successful. In 1999 a special amendment to the Sale of Liquor Act was created that exempts hosts from the requirement to have a licence if they serve, sell or supply liquor for no more than ten guests (@home New Zealand, 2008).

According to the domestic law of different countries, there may also be special regulations for running accommodation institutions, including rural guest rooms and agritourism farms. For tourist enterprises such as hotels, guest houses, tourist camps, agritourism farms and other institutions offering places to sleep, there may be regulations specifying the hygienic-sanitary requirements. In agritourism farms fire regulations for dwelling houses and farms are usually applicable. Dwelling houses and service houses as well as other structures and areas, such as campsites or recreational areas, should be designed, used and maintained in a way that prevents a fire.

Many of the regulations are very complex. They vary depending on the type of enterprise operated and where it is located as well as many other factors specific to particular activities; however, they should be carefully examined and followed by agritourism entrepreneurs before starting and while operating an agritourism business. It is appropriate to contact agencies for specific information on fees and to obtain the most current regulations and explanations of them (Agritourism in Focus, 2005).

Case Studies

1. Check which legal acts and regulations in your country and region may pertain to agritourist activity.
2. Check whether the law in your country imposes the obligation of legalization of economic activity and whether it defines forms of this activity.
3. Check in what way the law in your country defines a farm and what advantages and problems result from the fact in relation to agritourist activity.

7 The Organization of an Agritourist Farm

Types and Definitions of Farms

In the history of agriculture in Europe, a farm was a land property, comprising land together with buildings standing on it, designed generally to provide food for the family living on the farm. A farm was very closely related to the rural household. It was usually a multipurpose farm cultivating plants and breeding animals, as well as processing food. In the course of time these farms developed, as a consequence of which different forms of farms appeared. In the New World a farm is sometimes defined differently; in extreme cases the term refers to a separate area of land on which agricultural production takes place.

At present we may distinguish traditional farms, farms focused on plant production (crop farms), horticultural farms, orchard farms and animal breeding farms. These may also be divided in terms of the type and volume of production. A traditional farm is involved in plant and animal production and food processing. A family living on the farm usually owns it.

A farm is a production entity in agriculture, equipped with the necessary factors of production: land, buildings and facilities. In Europe, Asia and Africa farms are mostly family farms. In North America a separate term is used, i.e. a ranch, which refers to a farm breeding beef cattle. Agricultural production run on the farm is frequently only one form of household activity. Farms frequently expand their operations to include other branches of economic activity. Here agritourist activity is an option (Fig. 7.1).

The agritourist farm derives from the agricultural farm with business operations extended to include agritourist activity. The most frequent reason why an agricultural farm starts agritourist business is to increase income. Hence, before discussing the organization of an agritourist farm, it is necessary to discuss briefly the organization of a classic farm.

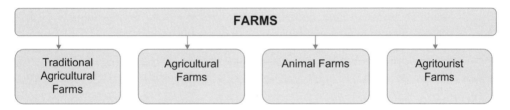

Fig. 7.1. Types of farms.

The Agricultural Farm

The farm is a basic production enterprise in agriculture. It consists of an area of land and other means of production such as buildings, appliances and tools. It has resources of human labour and tractive force. The whole of its activity is managed by a farmer or farmers. What distinguishes the farm from other production units is the fact that the soil is the most important means of production and that it naturally reproduces a number of necessary means. Usually the farm produces plant and animal products, and sometimes it also possesses a food-processing plant. The farm may be owned by a farmer, it may be managed by a tenant or it may be part of a company, cooperative society or even association.

In the farm two spheres can be distinguished. The technical-production sphere comprises the processing of some material goods into others. However, the transformation is dependent on biophysical constraints. The economic sphere specifies the production process with allowance for relations between prices and products as well as the financial value added during the production process.

The farm property

The land and means of production are the farm property. The property is the foundation from which production tasks are performed. The farm property components can be divided into fixed and variable assets. The fixed assets depreciate over time in numerous production processes and gradually lose their value in the production of new products and services. The variable assets are depleted in a single production process and lose their value in a single production process. In accounting practice, the division into fixed and variable assets is determined by individual regulations, which precisely specify into which group particular assets should be classified. Fixed assets include the following:

- Buildings, structures, equipment;
- Irrigation and drainage equipment;
- Machines, devices, tools, means of transport;
- Perennial plantations, e.g. hop gardens, willow and asparagus plantations;
- Office and other equipment, e.g. typewriters, computing machines, fire equipment;
- Base herd.

Variable assets include materials, feed, fertilizers, fuel, young animals (piglets, calves, fattened cattle and pigs), cash and low-value livestock.

Land

Farmland is limited in every country. The limitation results from the qualities of land, namely its spatial character, its immovability and the inability to create more land. These specific characteristics of land are important because the existing land derives from a natural endowment plus the work of many generations in cutting forests, stumping, levelling, irrigating and draining.

The hectare is the basic unit of land measurement in Europe. The land area does not specify the quality and usable value of the land. The most significant element determining the use value of land is the soil, which is characterized by the fact that it can provide plants with water, nutrients and air. The farming and use value of soil is classified in a scientific manner. In different countries different soil types are distinguished. Very frequently in the same farm there are soils of different types, so it is important for analysts and entrepreneurs to look at individual soil types as well as index values for an entire landholding. There is an internationally accepted classification of soils and their suitability for plant production.

The soil classes and the average soil index value do not give a complete view of the use value of land. The farming value of land can best be evaluated if soil and climatic conditions are taken into consideration simultaneously. Sandy soil in a humid climate where yearly precipitation reaches or exceeds 1500 mm has a much higher use value than the same soil in a dry climate where the yearly precipitation is under 400 mm.

Production structure of the farm

The structure of a production organization may be very complicated – we speak then of a traditional or multipurpose farm, or very simply we speak of a simplified or specialist farm. A traditional farm is one on which various plant or animal products as well as the products of the agricultural and food industry have been made for long periods. Usually the farms are oriented towards self-consumption of the products they have made. Specialist farms are those that have chosen one direction of production. They deliver large, homogeneous batches of products to the market. Diversified farms are those that have abandoned narrow specialization in order to avoid economic risk and run two or three directions of production on a high level. It is possible to distinguish activities within a farm. We can distinguish production activities, including crop production, livestock production, horticulture and market gardening, processing activities, service activities and general management activities. Because there is a large diversity of production activities, production activities can be separated into components, as is done in Fig. 7.2.

The organizational structure of the farm shown in Fig. 7.2 is very complicated. In practice the organizational structure of real farms is usually simpler.

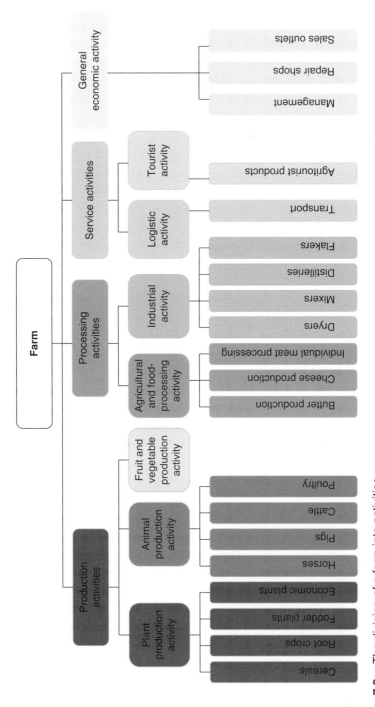

Fig. 7.2. The division of a farm into activities.

The production structure specifies the type of a farm. The farm may involve only commodity plant production (a cropping farm) or only animal production (dairy farm, pig farm, etc.). Farms may also be multipurpose. If a livestock farm is very large and involves cattle or sheep grazing, it is defined as a ranch. A farm may also deal with fruit production – then we speak of an orchard. The knowledge of farm management is very wide. This textbook does not present these issues; however, when discussing problems concerning agritourist farms it is necessary to refer briefly to this knowledge. Below is presented a brief description of the crop and livestock production activities in a multipurpose farm.

Crop production

The aim of crop production activity is producing plant products (fodder) for further processing on this farm and products ready to be sold to food enterprises, on the wholesale market, in a street market or in direct sales to the final consumer. Agricultural land plays the most important role on a farm; therefore we shall discuss its division and further describe it. It can be divided into:

- Arable land;
- Permanent grassland – meadows and natural pastures;
- Vegetable garden, hothouse grounds;
- Orchards, nurseries, permanent plantations;
- Underwater grounds.

Crop production activity can be characterized in many ways. One common technique is to compare the structure of agricultural land use and its productivity in the main crop. An illustrative allocation of agricultural land use and its productivity for a particular farm of 128.6 ha is presented in Table 7.1. This table provides information about the structure of crop production. For a full

Table 7.1. An example structure of using land and its productivity.

Crop	Area (ha)	Area (%)	Yield/ha (dt)	Harvest (dt)	Crop by-product	Crop by-product (dt)	Harvest by-product (dt)
Wheat	9.80	7.62	55	539	Straw	55	539
Barley	10.40	8.09	45	468	Straw	45	468
Silage maize	45.20	35.16	500	22,600		0	0
Sugarbeet	5.50	4.28	400	2200	Fresh pulp	200	1100
Rape	23.20	18.05	25	580		0	0
Lucerne	12.85	10.00	400	5140			0
Arable land	107.00	83.20					
Meadow (hay)	16.40	12.76	40	656			
Pasture	5.20	4.05	450	2340			
Area of agricultural land	128.60	100.00					

dt: decitonne (1 dt = 100 kg)

Table 7.2. Animal population, increase and decrease on a farm in a year.

Animal production groups	Population at the beginning of the year	Increase during the year	Decrease during the year	Population at the end of the year
Cows	200	34	34	200
Pregnant cows	26	42	42	26
Heifers aged 1–2 years	64	77	76	65
Heifers aged 0.5–1 year	35	80	79	36
Heifer calves aged under 0.5 year	45	84	83	46
Bull calves aged under 0.5 year	45	86	89	42

description of crop production activity, it is also necessary to specify the machinery and tractive force the farm is equipped with, the consumption of organic and mineral fertilizers and chemicals as well as the kind and number of services in crop production. In the last phase of the description of crop activity we specify the production technology for individual crops and the work necessary to produce them.

Livestock production activity

The aim of this activity is to produce animal products for further processing or breeding animals for further reproduction.

The fixed assets of the livestock production activity include the basic herd, buildings for livestock, structures, silos, manure disposal facilities and other equipment and means of transportation. The variable assets include other livestock and such materials as fodder, mineral additives, drugs, etc.

The animal population is registered according to species and age groups, production groups and others. The population is documented with the actual current number of animals or average population of animals in the recent period. The average population is worked out on the basis of the animal register at the beginning and at the end of the period. Table 7.2 includes an example of a description of an animal population. Livestock productivity is expressed in the description of the activity. It is expressed by:

- Annual milk yield;
- Daily increase (in grams or kilograms);
- Reproduction index;
- The final body mass of fattened animals;
- The production of wool per head per year;
- Egg production percentage.

The process of crop production and livestock production is very complicated. Farm management specialists deal with the integration of the technological and economic aspects of this process. This textbook is dedicated to the economics analysis of agritourism and it does not discuss the fundamentals of farm management, which have been described in detail in the literature on the subject.[1]

The Agritourist Farm

A farm that has introduced a new activity, agritourism, into its structure is called an agritourist farm. Besides crop and animal products the agritourist farm also provides agritourist products and services. Agricultural activities and the tourist activity can be independent of each other on a farm. In this situation there are no connections or flows between them. For example, a farm specializing in the production of cereals is situated in an attractive area and therefore it also provides services for tourists.

The activities of a farm may also be interdependent; however, usually the agricultural activities are partially or completely subordinated to tourism; for example, tourists use the food produced on the farm. The extent of subordination of the agricultural activities to the agritourist activity can vary.

In a 'purely' agricultural farm, 100% of the income comes from crop and animal production and from agricultural and food processing. On an agritourist farm, a certain part of the income comes from agritourism and the other part from agricultural production. Agricultural production may also be completely subordinated to agritourism and then 100% of the income comes from this activity. The index of the share of income from agritourism determines the importance of agritourism for the farm. The larger the part of the income that comes from agritourism, the more important it is for the farm.

The share of agritourism-related income in the total income from the farm varies considerably. According to a survey conducted in the Wielkopolska region of Poland in 2000, the share of agritourism-related income in the total income of farms was slight. Most frequently it ranged from 6 to 10% (21.3% analysed entities) and 11 to 20% (17.5% analysed entities). Only in 8.7% of studied farms did income from tourism exceed 50% of the total income (Table 7.3).

The principles of organization of an agritourist farm will be discussed in detail in Chapter 10 on agritourist services and products as well as in Part III, which describes examples of agritourist farms around the world.

The Agricultural Farm vs the Agritourist Farm (Figs 7.3 and 7.4)

A question arises as to whether there are any differences between the agricultural farm and the agritourist farm, and, if so, what are they? At the outset, it is necessary to emphasize that an agritourist farm must have a 'tourist' infrastructure that enables the reception of guests, whereas an agricultural farm, especially one specializing in crop production, does not even need to have dwelling houses

[1] Examples of textbooks dedicated to farm management are Shadbolt and Martin (2005) and Kay *et al.* (2007).

Table 7.3. The share of agritourism-related income in the total income of farms in the Wielkopolska region of Poland in 2000.

The share of income from agritourism/rural tourism in total income (%)	No. of farms/ responses	In % farms total
Below 1	14	7.7
1–2	12	6.6
3–5	27	14.8
6–10	39	21.3
11–20	32	17.5
21–30	22	12.0
31–50	18	9.8
50–75	9	4.9
Over 75	7	3.8
No response	3	1.6
Total:	183	100.0

The study covered all agritourist fams and rural tourism accommodation facilities in the Wielko-polska province in 2000, which agreed to fill in the survey questionnaire.

Fig. 7.3. View of the yard of a dairy farm in Manitoba, Canada (photo by M. Sznajder).

because the farmer may live beyond the limits of the farm. However, this situation is rather rare in Europe.

Agricultural farms and agritourist farms differ in the rules of running and managing them and frequently also in the production structure. In particular, this concerns the use of the land, the intensity of production, the management of

Fig. 7.4. View of an agritourist farm in Poland: notice how the surroundings improve dramatically when moving from agriculture to agritourism (photo by M. Sznajder).

work resources and environmental protection. The aim of running agricultural farms is usually long-run maximization of profit by intensive management. In view of this fact, agricultural farms try to use as much land as possible for crop or livestock production, which guarantees the highest profit, whereas agritourist farms frequently exclude even large areas from this production and instead use them for extra structures, car parks, lawns and flower beds, which are supposed to beautify the agritourist space, and squares for sports or recreational activity. The principles of running agricultural farms are usually different and sometimes they even stand in opposition to the rules of running agritourist farms. For example, agricultural farms aim at specialization in production and concentrate on one or a few directions of production, whereas for agritourist farms varying the production structure is a better solution because they thus become more attractive to tourists. Agricultural farms aim at intensification of production, which consists in, for example, using high amounts of fertilizers and pesticides, concentrating high numbers of animals per hectare and their high yield. The limits of intensification are determined either by marginal analysis or by legal regulations that specify the maximum concentration of animals per hectare or the maximum amount of fertilizers to be spread. There are also farms that seek to maximize their income by organic production techniques. Such farms expect higher prices for their products. Agritourist farms usually aim for low levels of farming intensity or for organic production because these farming systems are more compatible with the needs of agritourism.

On the other hand, agricultural farms usually aim at the introduction of technological progress. In the last 50 years production technologies have changed radically. The aim of technical advance was maximization of work efficiency. Modern technologies of production in agriculture are not very attractive to tourists. They are almost identical in all countries of the world, not only in the highly developed countries but also more and more frequently in the poorer countries as well. Besides, modern technologies of production make it impossible for tourists to participate directly in the production process, because they require that qualified labour should be employed, especially to operate machines. For agritourism old technologies of production are very important, those that were used until the end of the 1960s and which varied considerably in different countries or even regions of countries. In fact, agritourist farms accept technological progress only for more effective tourist service.

Agritourist farms also have a different approach to the market for their products. Modern markets are indispensable to conventional farms, whereas agritourist farms treat tourists as a serious market for their products. Hence the structure of crop and also even livestock production is at least partly subordinated to this approach.

The relationship with the country and its community is also perceived differently in these two types of farms. On agricultural farms, especially modern ones, production has a weaker and weaker relationship with the country and its community. More and more frequently, the country is not a market for the farm and the farm does not provide employment for rural people. The relationship usually concerns cooperation with other agricultural farms (territorial integration) such as, for example, common use of more expensive agricultural machines or help with some specific work. The country and its architecture and community are a special attraction to tourists, especially if the country offers its local tradition, culture and dialect – that is, everything that constitutes its individuality. There are multiple relationships between the country and agritourist farms, which enable them to develop and function better. Table 7.4 compares the most crucial differences in the principles of managing an agricultural farm and an agritourist farm.

Relationships between the Agricultural and Tourist Activity in Agritourist Farms

In view of the differences in the management practices of an agricultural farm and an agritourist farm a question arises: what is the relationship between the agricultural activities and the agritourist activity? There may be various types of relationships: complementary, supplementary, competitive and even antagonistic.

A complementary relationship consists in the fact that both activities constitute a deliberate structural entity and an increase in the importance of the agritourist activity causes an increase of the agricultural activity and vice versa. If a farm receives more guests, it needs more agricultural products that could be used as food. Apart from the external possibilities of sales of agricultural products, a new ready market opens. Guests staying on a farm often buy even those products the farmer could not sell otherwise. This relationship may also work in the

Table 7.4. A comparison of the fundamental principles of management of agricultural farms and agritourist farms.

The principle of management concerning:	Agricultural farm	Agritourist farm
Use of land	Maximization of agricultural use of land	Partial excluding of land from agricultural use, using it for agritourist purposes (buildings, car parks, squares, etc.)
Production structures	Only agricultural, breeding and possibly processing activities. Specialization of production, increasing the scale of production	The agricultural and agritourist activities in various proportions to each other. Subordination of the structure of agricultural activity to agritourism. Diversification, balanced development of farms
Work organization	Maximization of work efficiency, mechanization and automation	Organizing production processes in such a way that they are 'spectacular' or even give tourists a chance to participate in them. This requires high work investment. Thus work efficiency is not the leading principle
Investment level	Emphasis on production intensity, and investment in modern technology, especially fertilizers and pesticides, in order to maximize the profit from agricultural production	Emphasis on extensive production and environment protection. Moderate use of fertilizers and pesticides. Optimization of the quality of agritourist space
Market for agricultural products	Ready market beyond the agricultural farm, mainly contracting, purchasing	Using the products on one's own farm for the purposes of agritourism, excess sold on the market
Sources of income and their importance	All income is generated from crop and livestock production and possibly from food processing	Income comes from two sources, agricultural production and agritourist activity, or only from agritourist activity. The importance of agritourism for the farm depends on its share in the total income
The country	Agricultural production is done in rural areas, but rusticity does not have much in common with the production standard on a farm	The country facilitates running an agritourist business. There is a possibility and necessity to use folk traditions, for example (ethnography)

opposite direction. For example, if the farmer had a larger strawberry plantation where tourists could pick the fruit, he would have more guests.

Agritourist products may also be competitive in relation to certain agricultural activities. Competition may concern the use of all resources of the farm, i.e. land, work and capital. For example, a farmer growing commodity crops intends to develop agritourist activity. For this purpose he has to exclude part of the area of land from agricultural production and use it for agritourism, or part of the finances he has been directing to crop production so far must be directed to agritourist investments. Besides, he has to direct part of the work to tourist service.

There may also be an antagonistic relationship between the agricultural and agritourist activity, i.e. a situation in which one activity excludes the other. The most drastic example of antagonism between agritourism and agricultural production concerns integrated livestock production, especially pig herds. Agritourist activity near large pig, poultry or cattle farms is virtually excluded. Tourists usually do not accept this kind of production because of the offensive odour. All forms of agricultural production that are a source of unpleasant odours exclude agritourist activity. Poultry farms also are not a favourable element for the development of agritourism. However, on a multipurpose farm, where two or three pigs are kept, agritourists will accept them with pleasure. Also certain forms of crop production may not favour agritourism, especially those related to the use of pesticides. Agritourism may also exclude agricultural activity. In particular this concerns products that require strict limitation of access for veterinary and phytosanitary reasons.

Case Studies

1. Select and compare a 'typical' farm and an agritourist farm in terms of: (i) the amount and quality of land; (ii) fixed assets; (iii) plant production and animal production; and (iv) farming objectives.
2. Analyse the share in the total income of an agritourist farm of agritourism-related income. What are the consequences of this fact?
3. In view of Table 7.4, list basic differences in the management of a farm and an agritourist farm.
4. What actions would have to be taken to transform your 'typical' farm into an agritourist farm?

8 The Economics of Agritourist Enterprises

Introduction to Economic Analysis

Economic analysis of projects and economic enterprises is a very wide branch of knowledge and it considers a range of detailed problems. The literature discussing the economics of enterprises and production provides the necessary knowledge for running agritourist enterprises. Agritourist activity can be analysed with economic methods that are identical to those applied to analysing enterprises in other economic activities.[1] The methods of analysis are generic, but their interpretation is specific. This book will discuss only some techniques that are useful to explain and analyse selected economic problems of agritourist enterprises, namely:

- Resources and factors in agritourism;
- Analysis of outlays and effects;
- Listing categories of outlays, income and costs;
- Costing in case of production costs of agritourist products and services;
- Financial analysis of economic activity;
- Break-even point in agritourist activity;
- Risk in agritourism.

The importance of economic analysis for an agritourist enterprise is enormous. Above all, it gives a possibility to find if the business brings positive or negative financial results. If an agritourist activity brings profit, we further think how this profit could be increased. If the activity brings a loss, the aim of the analysis is to determine its causes and find a way to eliminate future losses. In such economic analysis we focus on events that have taken place. It is an *ex post* analysis. Results cannot be changed because they have already occurred.

---

[1] This book is not a lecture on the theory of economics. As a textbook in a branch of economics, it gives only the range of material of the theory of economics that is necessary to highlight specific problems of the economics of agritourism.

On the basis of an analysis of previous years it is possible to draw far-reaching conclusions.

In addition to *ex ante* and *ex post* analysis, it is appropriate to analyse where a business is at any point in time. This suggests daily monitoring of physical, organizational and financial performance indicators. Computer programs that have been developed make possible a current economic analysis on a daily basis. At every stage of business activity it is possible to check if activity is ahead of or behind expectations. The method dealing with ongoing control of income and expenses is called controlling.

It is also possible to make a forward-looking economic analysis of an agritourist business (*ex ante* analysis) based on projections of future business results. The heart of *ex ante* analysis involves forecasting and planning. This includes making a budget and drawing up a plan of future activity in order to achieve the set goal. Three basic scenarios of planning can be distinguished: optimistic, most probable and pessimistic.

The pessimistic option predicts that events may be unfavourable. If the pessimistic option shows possible profit anyway, there is a high degree of certainty that we have made the right decision. In planning we should tend to consider the pessimistic option, but it can make one too cautious.

The other extreme is the optimistic option. We assume that there will be favourable conditions for implementation and everything will develop well. The optimistic option fuels enthusiasm and expectations for economic success, but many an enterprise that planned its development assuming only the optimistic option has been a failure.

The third option of planning concerns the most probable situation. The realistic option is the one we are most likely to realize. Good planning is facilitated by economic forecasting.

Accounting data are of fundamental importance for economic analysis. Accounting documentation is extremely important and should be used for both tax and analytical purposes. Accounting sources are a mine of materials, which should be appropriately used. Accounting records are usually justified by the duty to settle accounts with the tax office as part of the business activity, or to settle accounts with a partner, shareholders or members. Frequently, use of the accounting records is limited to the calculation of certain key pieces of information, e.g. the amount of tax that must be paid. Frequently the other accounting data are not used. However, accounting data may be used to help manage an enterprise. Therefore more advanced financial analyses are necessary. The reports that are prepared for the tax office are made for different purposes from those for an analysis made for management purposes. Financial analysis useful for managers of an enterprise may, for example, include estimates of the ratio of turnover to assets and estimates of capital adequacy and liquidity.

Categories and Definitions of Costs

The economic activity of an enterprise is characterized by the categories of investment, income and expenses. The categories of income, investment and

costs are precisely defined in a particular country's tax law related to accounting. The investment is all non-material and material assets that are invested in
the production of agritourist products and services. There is a difference between
investment and cost. Investment is expressed in terms of asset values. Costs are
expenditures relating to that investment. Costs can be divided into property and
production costs. Property costs are generated in consequence of owning the
asset, independent of the level of business activity. Production costs are additional costs associated with the production of a product or service. The differences between these costs are significant. If someone has an asset, for example
a farm, which does not produce any products or services, the owner has to bear
some costs in spite of the fact that there is no production. It is necessary to pay
property taxes and maintain the asset. Thus, property costs are the costs that
must be borne despite the fact that the property does not produce anything.
Fixed costs have special importance in agritourism. Maintenance of the agritourist infrastructure generates high fixed costs. In particular, there is usually a low
season, for example winter, when there are no tourists at all. Thus, production
costs are all those costs borne to manufacture agritourist products and provide
agritourist services.

    We may also distinguish fixed and variable costs. They refer to the scale
(volume of activity). Variable costs increase with an increase in scale. The more
tourists, the higher the costs connected with their service, e.g. catering or energy.
Fixed costs do not depend on the scale of activity. Irrespective of the number of
tourists, depreciation of a building is always the same. Fixed costs decrease with
an increase in the scale of activity per unit. The more tourists we serve, the lower
the depreciation cost per person.

    Costs can also be divided into direct costs and indirect costs. Specifying
direct and indirect costs is very important when various agritourist products are
manufactured and various agritourist services are provided as part of the agritourist activity. Indirect costs are borne for all activity of an agritourist enterprise.
Direct costs are the costs that can be directly attributed to a specific agritourist
product or service. Indirect costs are the costs that cannot be directly attributed
to a specific agritourist product or service.

    Apart from the division into direct and indirect costs, costs can also be clas-
sified into groups. The classification of costs is potentially very complicated,
because the products are diverse. In fact, for each product there are specific
costs. For example, for the agritourist service of horse riding the following types
of costs can be distinguished: wages and their derivatives for a horse service,
wages for the instructor and trainer, equipment costs, fodder costs, depreciation,
breeding costs, animal purchase, maintenance costs and others.

    Another very important category is opportunity costs, i.e. lost opportunities
costs (lost profit). An analysis made by means of this category makes it possible
to decide whether the choice and financing of a specific option from the set of
many different activities was right. For example, the financial assets could have
been deposited in a bank account with an interest rate of 3%, whereas investing them in the business activity brought a profit of 6%. The bank interest is a
missed opportunity. If we allow for the opportunity cost, the net profit amounted
to only 3%. It is important to remember that the opportunity cost is not

necessarily the bank deposit rate. It is the value of the resources in their best alternative use. The analysis with allowance for opportunity costs is of central importance when making financial decisions.

In economic analyses an important element is the study of cost structure. Cost structure, i.e. the share of individual types of costs in the total cost, is characteristic for many products and services.

Calculating Enterprise Costs

The calculation of costs of individual services in a multipurpose agritourist enterprise requires a number of methodological explanations. Each activity has its own costs and income. However, certain farm activities do not bring direct income, but only generate costs. For example, feed production does not bring income, but it generates costs. A similar situation occurs in the case of administrative activity. In spite of this, both feed production and administrative activity are necessary in an enterprise. The calculation of production costs in the activities generating income consists in allowing for the direct costs of these activities and the indirect costs, which originate in the activities that do not generate income.

For easier understanding of the rules of calculating production costs for an enterprise, it is necessary to familiarize oneself with the terms profit centre and cost centre. A profit centre is defined as a product or service centre that generates revenue and costs for an agritourist enterprise. A cost centre is defined as an activity centre that generates costs but does not generate revenue, as with the previously mentioned feed production and administrative activities.

Generally the income centre is at the same time the cost centre, although there are cost centres that do not generate income, as, for example, feed production and the administration department. An agritourist farm offers tourist services in the form of horse riding and accommodation with breakfast, as well as running agricultural operations – commercial plant production and production of forage for horses. The analysis of costs for this farm may include the following costs:

1. Agritourist services:
   • horse riding;
   • bed and breakfast.
2. Crop production:
   • horse feed;
   • commodity production.

On the farm there is also the administrative activity serving the agritourist activity and agricultural production.

The profit centres are: horse riding, bed and breakfast and commodity crop production. The cost centres are feed production, commercial plant production, accommodation with breakfast and administrative activity. Cost centres and income centres are specified depending on the nature of the enterprise and the needs of the analysis.

Summing up the considerations presented above, it may be stated that income and cost centres in the farm analysed are horse riding, accommodation with breakfast and commercial plant production. Sole cost centres are plant production for forage and administration.

The aim of the calculation is to determine – based on bookkeeping records – costs and income of individual branches of production and services in the agritourist farm. Table 8.1 presents the yearly income and costs of all cost and income centres in thousand Euros on the basis of accounting data.

The horse-riding costs amounted to 200 thousand euros, while the income gained amounted to 400 thousand. The financial result for this activity was 200 thousand euros. Similarly the financial results for the other income and cost centres were calculated. It was found that horse riding provides a gross profit of 200 thousand euros, bed and breakfast 100 thousand euros and crop production 50 thousand euros. We notice that the feed production and administration generate only costs, which amount to 50 thousand euros and 200 thousand euros, respectively. The feed is only for horses. Therefore the costs of this activity are transferred to the horse riding, which is marked with an arrow in Table 8.1. After adding the feed costs the horse riding gross profit decreased to 150 thousand euros and the financial result of feed production equals zero.

Similarly administration costs need to be covered. These costs must be distributed between all profit centres according to some rationale, for example, direct labour costs or according to indirect costs. Such an operation is called

Table 8.1. The costs and revenue of individual centres on an agritourist enterprise (in thousand euros).

| Specification | Profit centres | | | Cost centres | | |
	Horse riding	Bed and breakfast	Commodity crop production	Feed production	Administration	Total
Costs according to invoices	−200	−100	−100	−50	−200	−650
Income according to invoices	400	200	150	0	0	750
Gross profit = income − costs	200	100	50	−50	−200	100
Adding the costs of feed production to horse riding	−50					
Gross profit of all cost centres	150	100	50	0	−200	100
Distribution of administration costs between profit centres	−100	−50	−50			
Net profit of all centres	50	50	0	0	0	100

the allocation of overheads to individual profit centres. The distribution in which we add the costs equally to all the profit centres is wrong. In the case under consideration the distribution of administration costs was done by percentage according to the proportion of direct costs. Horse riding generates 50% of direct costs of the whole enterprise and that is why 50% of administration costs, i.e. 100 thousand euros, was added to the horse-riding profit centre. Similarly 25% of administration costs, i.e. 50 thousand euros, was added to the bed and breakfast service and 50 thousand euros was also added to the commodity crop production. As a result of the calculations, the net profit from horse riding amounted to 50 thousand euros, 50 thousand euros from bed and breakfast and 0 euros from commodity crop production.

Prices, Information and Strategy

The prices of agritourist products and services have not only economic but also informative value. The economic value was discussed in the previous paragraph. The informative value has vital importance both for potential clients and for service providers. Through the price the potential client is informed about the standard of the offered service and, on the other hand, the agritourist farm expects to receive a certain type of clients. The agritourist farms that are not ready to receive certain groups of clients use the price to ensure the inflow of clients from the desired 'segment'. The lowest prices are frequently chosen by young people. Many farmers are afraid of this group of clients. A high price simultaneously means a lower inflow of clients, but also less volume. Some farmers prefer to have fewer tourists for a higher price rather than more tourists for a lower price.

Some clients reject the services under a certain price in advance because they expect that low prices would not guarantee them an acceptable standard of services. Therefore they choose to stay on an agritourist farm with higher prices. Others look for moderate prices and yet others for extremely low prices. A beautiful room with a nice view, a separate bathroom and toilet, swimming pool, safe car park, dinner with the hosts at a table full of food and a neat garden cost money. The price must cover the costs and generate a margin. In New Zealand staying on an agritourist farm in luxurious conditions can cost as much as 1000 euros per person per day. The opposite is a bed for 20 euros in a collective room with a shared bathroom and washbasin, which are frequently of doubtful cleanness, without secured property, especially at night, without a meal or a car park. In Poland tourists can find fairly good-quality accommodation even at the price of 50 euros per person per day.

The price level combined with the service quality signals the pricing strategy. Kotler (1999) distinguished as many as nine different pricing strategies that fit the conditions of agritourist farms. The strategies are presented in Table 8.2. When using certain strategies it is necessary to realize the economic consequences. For example, the extortion strategy, which supposedly works in Parisian hotels, has no chance of success for Polish agritourist farms. Here the super-bargain strategy is probably viable.

Table 8.2.   Pricing strategies depending on the quality of the service or product offered according to Kotler (1999).

	Price		
Quality	High	Moderate	Low
High	Highest price strategy	High value strategy	Super-bargain strategy
Moderate	Overload strategy	Moderate value strategy	Good bargain strategy
Low	Extortion strategy	Apparent money-saving strategy	Money-saving strategy

Break-even Analysis

An investor making a decision wonders if and when the capital invested in production or a service will pay off. For example, a farmer planning to start an agritourist business on his farm wonders how many guests he will have to receive for the assets invested in the business to pay off. The answer to this question is given by an analysis of the break-even point, which is made in the coordinate system XY. Axis Y is the scale of the investment and income expressed in the monetary value, e.g. in thousand euros. Axis X is the scale of the enterprise in physical units or values. In agritourist activity the scale unit could be a tourist-day. The break-even point specifies the number of products necessary to cover the costs incurred.

For example, a farmer invested 500 thousand euros in the modernization and adaptation of an existing agricultural farm for agritourist purposes. The preparation of ten double bedrooms with bathrooms consumed 200 thousand euros, the rebuilding and modernization of the horse stable and the purchase of five horses cost 200 thousand euros and the modernization of the yard and recreational garden cost 100 thousand euros. Having made such a big investment or generated an annual interest cost of five thousand euros, the farmer wants to determine the number of tourist-days necessary for the invested assets to pay off. The maximum yearly number of tourist-days on this farm may be 365 days × 20 beds, i.e. 7300 tourist-days. However, usually the average use of the rooms does not exceed 40%, which means that in reality he can count on 2920 tourist-days a year.

Let us check the break-even point for the investment in question assuming that the unit cost of an all-day stay of an agritourist on this farm (bed + board + 3 hours of horse riding) is 80% of the revenue received. The farmer may apply various pricing strategies: the first strategy of a high price (150 euros/person/day), the second strategy of a moderate price (120 euros/person/day) and the third strategy of a low price (90 euros/person/day).

Figure 8.1 shows the net revenue (excluding interest costs) depending on the applied pricing strategy. With the high price policy the agritourist farm in question will reach the break-even point with less than 5% occupancy. In the case of the moderate price the break-even point will be exceeded with approximately

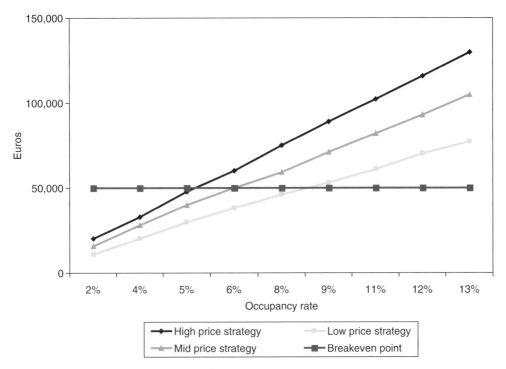

Fig. 8.1. An analysis of break-even points of an agritourist investment depending on the pricing strategy.

6% occupancy. If the farmer applies the low price strategy, the break-even point will be reached with less than 8% occupancy. The fact that these investments break even with such low levels of occupancy is because the analysis is considering only the marginal cost of the enterprise and does not consider the value of the entire asset.

Apart from the number of tourist-days sold, the farmer is also interested in the period of time necessary to have the invested assets completely paid off. The point is to specify the time spans in which the aforementioned break-even points can be reached. The farmer considers three options of farm use during the year: the optimistic one assumes the use of 50%, the most realistic one 40%, and the pessimistic one 30% (see Table 8.3 for each pricing category). The calculations show that in the best case the investment will be paid off after 2 years and in the worst after 23 years. This suggests extreme caution if the low price strategy is all the market will bear. The farmer can also consider lowering the costs of the offered services, which would also contribute to a quicker payback period. By lowering the cost of the service, the profit increases. On the other hand, there is a danger that the quality of the offered service will deteriorate dramatically and sales targets will not be achieved.

Table 8.3.   The time (years) necessary to reach the break-even point depending on
the pricing strategy and the accommodation utilization rates.

	Time in years necessary for the investment to be repaid		
Pricing strategy	50%	40%	30%
High price	2.00	2.50	3.25
Moderate price	3.50	4.50	6.00
Low price	13.75	17.00	22.75

Financial Liquidity

The main sources of financing the agritourist activity are income from the sale of
products and services and the funds borrowed (usually from a bank). The enterprise also has a possibility to finance the activity from other sources. The ability
to settle liabilities is defined as financial liquidity. The analysis of financial liquidity shows the ability of an agritourist enterprise to settle its payment liabilities in
different periods of time – year, month, week or even day. As an example we
shall analyse the financial liquidity of an agritourist enterprise that decided to
incur a debt of 340 thousand euros at an annual interest rate of 10%. To service
the debt an equal payment of principal and interest of 20 thousand euros quar-
terly was agreed on. In this system the debt will be paid off after 6 years. The
agritourist enterprise generates income from two types of activity: the first is
defined as 'contact with farm animals'. This involves one-day school excursions.
Another form is a holiday service, which includes accommodation and meals,
and it is offered to tourists coming to stay for a longer period of time because of
the attractive geographical location of the farm. The sale of the educational service is consistent during the school year and it generates an income of 5 thousand euros a month. The accommodation and food and beverage service for
tourists has a seasonal character. The income from this activity amounted to 5
thousand euros in January, 5 thousand euros in February, 5 thousand euros in
April, 15 thousand euros in May, 15 thousand euros in June, 40 thousand euros
in July and August, 15 thousand euros in September and 5 thousand euros in
October. In March, November and December no income was gained from the
activity. The farm has fixed costs of 5 thousand euros a month and variable costs
depending on the number of tourists.

   Figure 8.2 shows the income (positive cash flow) and expenses (negative
cash flow) on the agritourist farm under consideration during the year. Direct costs
rise considerably from May to August, when the number of tourists increases.
The costs also rise in March, June, September and December, when the payment of an instalment of the investment credit is due.

   Figure 8.3 shows the financial liquidity of the agritourist farm under consideration assuming zero starting cash and: no overdraft facility; a 10,000 euro overdraft facility; and a 20,000 euro overdraft facility. Interest rate on outstanding

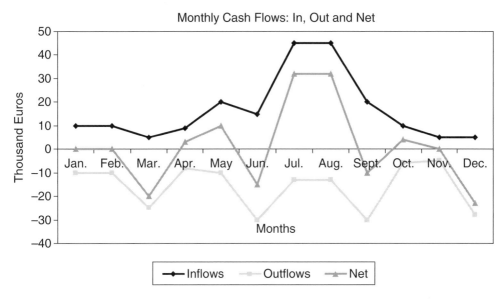

Fig. 8.2. The costs borne and income gained during the year on an illustrative agritourist farm.

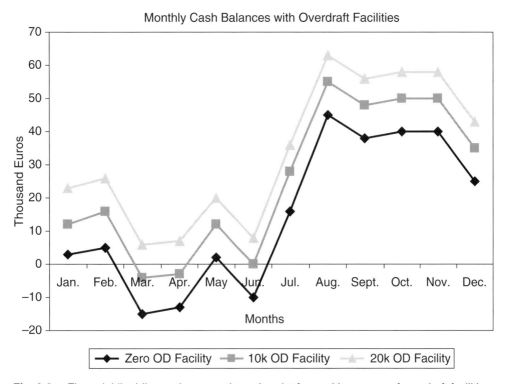

Fig. 8.3. Financial liquidity on the example agritourist farm with a range of overdraft facilities.

liabilities is 1% per month. This assumes that outstanding balances incur an interest cost of 1% per month. A key issue is what happens to the positive cash balances during the latter part of the year. If these are protected, this can enhance liquidity the next year. If they are spent, liquidity can be harmed in the subsequent year.

If a farm is not able to settle its financial liabilities without delay, it does not have financial liquidity and lacks liquid financial assets. This does not mean it is bankrupt. After all, it has significant assets. Many companies are not able to settle their liabilities without delay. In order to ensure financial liquidity, enterprises aid themselves with a bank overdraft facility. The life of this facility may extend over many years.

Investment Analysis

Agritourist entities investing in agritourist projects need to be able to evaluate the quality of an investment proposal. The investor needs guidance as to when an investment proposal should be accepted or rejected. The investor also needs to be able to choose between competing projects. The primary way to evaluate a project for which we can monetize all impacts is to calculate the net present value (NPV) of a project. The net present value calculation determines the profitability of an investment given that expenditures and revenues occur at different points in time. The NPV is defined as:

$$NPV = \sum_{t=0}^{n} \frac{(B_t - C_t)}{(1+r)^t}$$

where t = time, B = benefits, C = costs, r = interest rate.

When the NPV is positive a project should be considered as an investment possibility.

We can also calculate the benefit–cost ratio, which is the ratio of the net present value of benefits divided by the net present value of costs:

$$B/C = \frac{NPV(B)}{NPV(C)}$$

When the B/C ratio is greater than 1.0 a project should be considered as an investment possibility. With both the NPV and the B/C ratio, it is important to remember that the result is sensitive to the interest rate used. Hence it is always appropriate to do a sensitivity analysis that includes a higher interest rate.

The internal rate of return (IRR) is the interest rate r in the NPV equation that results in the NPV being equal to zero. When the IRR is greater than current and expected borrowing costs, a project should be considered as an investment possibility. Analysts need to be careful in using B/C ratios as problems can emerge with mutually exclusive projects and projects of different size. Similarly, when calculating the IRR, problems can emerge when there are mutually exclusive projects and when the project cash flows change their sign over time.

Table 8.4. Project one with 10% interest rate in euros.

Year	Costs	Benefits	Discounted costs	Discounted benefits
0	100,000	0	100,000	0
1	20,000	37,500	18,182[a]	34,091
2	20,300	38,250	16,777	31,612
3	20,605	39,015	15,480	29,313
4	20,914	39,795	14,284	27,181
5	21,227	40,591	13,180	25,204
6	21,546	41,403	12,162	23,371
7	21,869	42,231	11,222	21,671
8	22,197	43,076	10,355	20,095
9	22,530	43,937	9,555	18,634
10	22,868	44,816	8,817	17,278
			230,014	248,449
Indices	NPV			18,435
	B/C ratio			1.08
	IRR			13.96%

[a]Discounted costs for the first year are calculated as follows $\dfrac{20000}{(1+01)^1} = 18182$

Table 8.4 reports the analysis of an accommodation project that costs 100,000 euros and generates a revenue of 37,500 euros in its first year and costs of 20,000 euros. Each year the project revenue grows at a rate of 2%, while costs grow at 1.5% because of greater efficiencies. The net result is that the project has an NPV of 18,435 euros, a B/C ratio of 1.08 and an IRR of 13.96%.

Investment analysis provides the opportunity to compare the sensitivity of the project with alternative projections of benefits, costs and interest rates. It is also relatively straightforward to compare alternative projects. Table 8.5 presents the same project but using an interest rate of 15%. Not surprisingly, this results in the project having a negative NPV (–4106 euros) and a B/C ratio of less than 1 (0.98). At this interest rate the project is no longer viable.

The Risks of Agritourism

Like every business activity, agritourism has its risks. The risk in agritourism applies both to the entities providing agritourist services and the tourists who use them. For the supply party it involves either possible loss within the farm or loss resulting from events not related to the farm. Tourists arriving at a farm might unconsciously transfer animal and plant diseases or might cause an accident leading to the death of animals or damage to crops. Food-processing plants are in danger of pathogens, e.g. unwanted strains of bacteria. In order to minimize this risk many plants make galleries that enable observation of food processing

Table 8.5. Project two with 15% interest rate in euros.

Year	Costs	Benefits	Discounted costs	Discounted benefits
0	100,000	0	100,000	0
1	20,000	37,500	17,391	32,609
2	20,300	38,250	15,350	28,922
3	20,605	39,015	13,548	25,653
4	20,914	39,795	11,957	22,753
5	21,227	40,591	10,554	20,181
6	21,546	41,403	9,315	17,900
7	21,869	42,231	8,221	15,876
8	22,197	43,076	7,256	14,082
9	22,530	43,937	6,404	12,490
10	22,868	44,816	5,653	11,078
			205,649	201,543
Indices		NPV		−4,106
		B/C ratio		0.98
		IRR		13.96%

without contact. At the beginning of the 21st century, travel agencies with agri-tourist business had big financial losses caused by events that were completely independent of them. The losses were caused by foot-and-mouth disease in Europe, which completely paralysed the agritourist business for some time, as well as by terrorist attacks and wars. A rumour or unstable political situations may ruin many agritourist entities. For example, the emergence and rapid spread of foot-and-mouth disease in the United Kingdom is an example of the influence of unpredictable environmental phenomena on the development of agritourism. As Daneshkhu (2001) quotes in comparison with March 2000, in March 2001 an 11% fall in the total number of tourists was noted, including a fall in the number of tourists from North America by 16% and a drop in the number of guests from Europe by 9%. Simultaneously, instead of the 2.52 billion pounds' worth of income from tourism planned for the first quarter of 2001 (tourists' expenses in the United Kingdom), 2.36 billion pounds was earned.

The risk concerns the other parts of the farm property and delivered services as well. The risk rises from both conscious and unconscious tourist activity. Part of the property may be stolen, deliberately or undeliberately damaged or even fired. Tourists may even leave the farm without paying for the service.

The tourist using agritourist services is also exposed to the risk of health or safety hazard. These risks are the same as those in conventional tourism. However, sometimes the dangers are greater; the accidents caused by scared animals may be particularly dangerous and unattended small children may be the victims of serious accidents in contact with machines. Dogs biting, animal kicks or even butting are more and more frequent and serious. Besides, there is a risk of animal diseases such as brucellosis and tuberculosis, and recently also avian influenza.

People are increasingly more allergic to different antigens. The list of allergens gets longer and longer. A trip to tropical countries is related to a risk of contracting tropical diseases. In order to ensure health safety, it is necessary to be vaccinated and take anti-malaria medication. Vaccinations and drugs do not provide complete health safety. Amoeba is a dangerous blood parasite and, to avoid it, it is necessary to follow a certain preventive regime.

One needs to be cautious when entering a local community. Unfortunately, there are confidence tricksters, pickpockets and thieves just waiting to take advantage of tourists and thus preventive measures must be followed.

In New Zealand owners of farms are required to inform of possible dangers that may be faced by persons staying on their farms, including tourists. Any risk from the agritourist activity needs to be covered by liability insurance. Offering access to one's own land may be connected with accidents, even, for example, a broken leg due to slippery stairs or an uneven surface outside the house. In Canada standard liability insurance is 400 Canadian dollars annually. However, when more risky attractions are offered on the farm, such as horse riding, the cost of insurance increases considerably, to as much as 1400 dollars a year. Insuring some forms of agritourist activity is so expensive that farmers prefer not to have them in their offer. Farmers, especially those offering horse-riding recreation or participation in farm work on their farms, increasingly buy liability insurance.

Civil liability consists in the necessity to suffer the financial consequences of one's actions causing harm to others. Such a liability may result from the non-performance or inappropriate performance of a contract and causing harm to another person by an illicit act. Usually civil liability takes the form of damages, i.e. a certain sum of money corresponding to the actual value of suffered harm (e.g. damaged property) and lost benefits (e.g. lost income). Voluntary liability insurance is one of the ways to reduce the risk connected with civil liability in relation to guests staying at an agritourist farm.

Case Studies

1. Go back to Case Study no. 4 in Chapter 7 to the farm you intend to transform into an agritourist farm and determine the amount of necessary investment into its restructuring.

2. Estimate monthly income and expenditure of the new agritourist farm and prepare the annual cash flow.

3. Using the threshold point method, determine the number of months required to cover the costs of investments.

9 Cooperation of Agritourist Entities

Agritourist Regions

In most countries there are regions that are not predisposed towards agritourist business, where there are no agritourist farms and individual agritourist farms emerge only occasionally. There are also areas particularly suited to the agritourist activity. As a rule, in these regions more farms of the type emerge. So an agritourist region is an area where numerous agritourist farms function. A fragment of a map of a certain region (Fig. 9.1) shows an example where more than 20 agritourist farms function in the south-east, whereas in the areas situated north, east and west there are no agritourist farms at all.

Agritourist regions may comprise larger areas, for example several communes or districts. Mountain areas, where nearly all farms are capable of providing agritourist services, are a large agritourist region. Agritourist regions are also concentrated in areas of lake districts or some river valleys.

Types of agritourist regions

Agritourist regions can be divided into two groups. The first group is tourist regions with a single focus. In a tourist region with a single focus, all enterprises offer the same or very similar services and products. The opposite is diversified agritourist regions. This means that individual farms offer diversified different agritourist products and services. Homogeneous agritourist regions tend to cater for tourist holidaymakers who want to spend a longer period of time in one place, for example, 2 weeks. Diversified agritourist regions cater for mobile tourists, who want to use different products and services.

Both regions with a single focus and diversified regions have advantages and disadvantages from the point of view of both agritourism itself and a farmer providing agritourist services. We shall start considering the advantages and

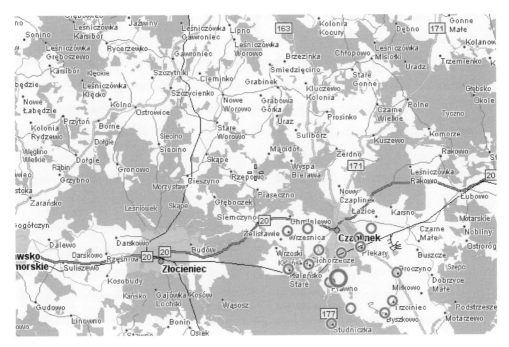

Fig. 9.1. A map of the agritourist region near Czaplinek, Poland (a fragment of Mapa Agroturystyczna woj. zachodniopomorskiego, 2003, reproduced with permission).

disadvantages in the regions with a single focus. Let us assume that somewhere by a beautiful river or lake there is a summer resort village. Tourists arrive in the village for a 2-week or 3-week holiday. All agritourist farms in the area offer the same products and services: bed, board and a holiday by the water. The advantage of regions with a single focus consists in the fact that clients know for sure that, if they arrive in a particular village, they will get the specific service they expect, no matter which farm they stay at. Farms in regions with a single focus can count on regular clients expecting summer holiday services. Summer resort villages are well known in specific city environs. Unfortunately, going on a summer holiday is possible only for a short period of the year in Northern Europe. Usually this does not start sooner than mid-May and does not last longer than mid-September. Then providing agritourist services comes to a standstill for a long period of time. Regions with a single focus can function all year round in a congenial climate, e.g. in Mediterranean countries.

If we consider the problem from the point of view of a farmer, who wants to prolong the period of providing agritourist services, regions with a single focus have a big weakness, because they do not offer products and services attracting tourists and holidaymakers in the low season. Another weakness of regions with a single focus is the possibility of fierce competition for clients between farms. If the number of tourists is relatively small, the competition between farms becomes even more intense, which is above all expressed by the lowering of prices for agritourist services. Until the end of the 20th century regions with a single focus

Fig. 9.2. Wine tourism in Barossa Valley, Australia, is a huge business: Miranda Cellars (photo by L. Przezbórska).

dominated. An interesting example of an agritourist region with a single focus is the Barossa Valley in Australia (Figs 9.2 and 9.3). The local community, thanks to cooperation, created a wine tourism region (see Box 9.1).

Diversified agritourist regions provide varied products and services. This means that each farm provides different products or services. Due to the diversification the farms in the region do not compete with each other directly, as is the case in regions with a single focus, but, on the contrary, there is a chance for cooperation and specialization. In diversified regions, above all, there is a chance to prolong the period of using agritourist services, and especially to develop weekend tourism. Of course, cooperation between farms in diversified regions is not automatic. It is created by associations or companies facilitating the development of agritourism. To sum up, it is possible to say that agritourist farms in diversified regions have, at least potentially, a definite advantage over regions with a single focus.

Integration in Agritourism

Integration is a term that means connecting business entities for the realization of common goals. There are various forms of integration. Complete integration takes place when the cooperating entities merge and form a new entity. Complete integration may also consist in the fact that the stronger entity incorporates the

Fig. 9.3. Wine tourism in Barossa Valley, Australia: there are more than 80 wine makers and 500 grape growers (photo by L. Przezbórska).

weaker entity or it may mean the joining of two or more equal partners. Integration may also mean cooperation of independent partners for the realization of mutual goals. There are three types of integration: horizontal, vertical and territorial.

Horizontal integration takes place when two or more business entities of similar production or service profiles cooperate with each other for the realization of a common goal. For example, two agritourist farms together purchase the means for client service. Vertical integration takes place when entities of the same branch but occupying different positions in the marketing link join together for the realization of common goals. For example, an agritourist farm cooperates with a travel agency and a transport company. Territorial integration takes place when entities not necessarily of the same branch but located in the same area cooperate with each other for realization of common goals. There are obvious benefits from horizontal, vertical and territorial integration.

The Position of a Single Agritourist Farm

The influence of an individual agritourist farm on potential clients is smaller than the influence of a group of farms. An individual farm has to act on its own to gain clients. Advertising and promotion expenses are necessary. These costs may be smaller as long as the farm has its own loyal clients who visit it every year. Clients' loyalty guarantees a regular income. However, loyalty in agritourism is limited in time. As years go by, clients change their taste or their conditions change. A farm catering for little

Box 9.1. Wine tourism in Barossa Valley

Wine tourism refers to a kind of tourism aimed at visiting wine regions as well as tasting, consuming or purchasing wine during the tour. Tourists can visit wineries to see how wine is made. Many wine regions have found it financially beneficial to promote such a tourism industry. Grape growers' associations and others working in the hospitality industry in wine regions have spent significant amounts of money over the years to promote such tourism and wine tourism plays an important role in advertising their products. Wine tourism has developed around the world, not only within the so-called Old World wine-producing countries, such as Spain, France or Italy, but also in the New World wine regions, including Australia, Argentina, South Africa and Chile (e.g. the Barossa Valley in Australia or the Mendoza Province in Argentina).

The Barossa Valley is one of the most significant and most famous wine-producing regions of South Australia (Barossa South Australia, 2008; Official site of the South Australian Commission, 2008). The valley is located 60 km north-east of Adelaide by the North Para River. There are also eight other distinct wine regions within an hour and a half of the Adelaide city centre, namely, Eden Valley, Adelaide Plains, Adelaide Hills, McLaren Vale, Southern Fleurieu, Currency Creek, Langhorne Creek and Clare Valley.

The Barossa Valley is approximately 13 km long and 14 km wide and is cut by the Barossa Valley Way, which is the major road connecting the main places on the valley floor of Nuriootpa, Tanunda, Rowland Flat and Lyndoch. The major towns of the Barossa Valley are generally recognized as being of German origin with long-standing traditions dating back to the 1840s when the first German and British settlers arrived in the area. The Barossa Valley has become a world-famous wine tourism destination, offering visits to wineries, vineyards and restaurants known to offer unique vintages, as well as organized wine tours and a variety of wine festivals. Every two years a week-long Vintage Festival draws visitors from all over the world, including a street parade, concerts, gourmet dining and other special events. There are more than 80 wine makers and 500 grape growers that contribute to the ongoing reputation of the region. Well-known names such as Jacobs Creek, Penfolds, Wolf Blass, Peter Lehmann, Saltram, Grant Burge, Torbreck and Seppelts are based in the Barossa Valley. However, although the region is overshadowed by the wine industry, there is also significant food production in the Barossa Valley, including dried fruits, pâté (a form of spreadable paste, usually made from a mixture of meats, including liver, additional fat, vegetables, herbs, spices, wine and other ingredients, and often served with toast as a starter), verjuice (a very acidic juice made by pressing unripe grapes, sometimes with an additive of lemon or sorrel juice, herbs or spices), ice cream and bakery products. Apart from the wine making, the valley is renowned for its diverse culture and boasts of a rich European history of home-made cooking as well as handicrafts. Each year the valley is visited by numerous tourists from all over the world.

children must allow for the fact that children grow and, when they reach a certain age, they will not want to visit it any more. Owing to the fact that individual farms are not able to bear the costs of promotion, joint action is one approach. Agritourist farms may form groups of farms, networks or agritourist associations.

Agritourist Groups

The fact that integrated farms can support one another is their advantage over individual farms. Cooperation and integration between agritourist entities take various forms, from simple to very advanced integration processes. The simplest form is informal joint actions, which are ad hoc actions if necessary. Joint actions are taken in order to facilitate the functioning of agritourist entities and to lower their costs. The costs resulting from the cooperation are borne in solidarity by all participants in proportion to the benefits resulting from the cooperation. Informal simple cooperation of agritourist entities may apply to common transport of equipment and tourists, organization of events together, etc. The cooperation is an element of territorial integration, because it is usually started by entities functioning in the same area.

Territorial cooperation may also be more formalized. A group of agritourist entities establishes a company or association for the realization of common goals. When forming superior structures, usually each farm retains its legal identity and is independent. The stimulus to developing cooperation within a farm group is the possibility of a considerable lowering of the costs of the agritourist services provided and increasing the income.

Agritourist entities situated in the same area may also start no cooperation. The reason why no cooperation starts is that other entities are treated as competitors, which must be rivalled and defeated. The entities applying the strategy of competition do not promote any cooperation. Also, weak agritourist entities do not usually start any cooperation as they are anxious about their future.

Agritourist Networks

Agritourist networks are a more advanced form of cooperation of agritourist entities. Within agritourist networks, franchising agritourist networks can be distinguished.

An agritourist network is a group of diversified entities (farms) that are not necessarily close to one another but which cooperate with one another so that one farm informs the tourist staying on it of the attractions of another farm belonging to the network. Networks of agritourist farms cater for a mobile tourist, travelling by car or bicycle. The entities associated in networks cannot offer identical products or services. Each of them should offer something unique. If farms in a network offered the same products and services, they would compete within the network because the tourist would not need to visit another farm offering the same profile of products and services. Farms forming an agritourist network can function on the basis of a mutual agreement or establish a more formal structure such as an association. In this case the association functions as a non-profit organization. By establishing an association the farms remain independent; however, they finance its activity. Associations are entrusted with various agreed tasks, for example, preparation of a shared brochure for all entities or analysing the stream of tourists, etc. Associations will be discussed in more detail in the next section.

An agritourist network is above all aimed at serving short-term clients who spend a few hours on one farm and then intend to visit another. Agritourist networks have good prospects due to the fact that there has emerged a group of mobile tourists who very much enjoy leaving their homes for a weekend. They get into their cars and travel into the unknown, frequently changing the places of their stay. If a mobile tourist visits at least one agritourist farm belonging to a network, there is a chance he will visit the other farms of the network.

An example is a group of farms in Yorkshire in the UK, which invested in producing a map of the network and published it in different versions. In the restaurant belonging to the agritourist network one can find on the tables colourful paper napkins with a map printed showing the location and attractions of the farms belonging to the network. The map comprises an area whose diameter is about 80 km and there are travel routes and staging posts marked on it. Each farm of the network offers different attractions. While waiting for a meal and looking at the map, it is possible to choose the place for another stopover.

A franchising agritourist network is virtually the opposite of an ordinary agritourist network. The difference is that agritourist farms offer the same products and services of the same standard. The entities belonging to the network use the same logo and decorations. Franchising is a specific form of licensing services. Usually the owner of the licence is a company. A farm belonging to a franchising network is a separate business entity that has to strictly conform to the conditions of the licence. The company (person) giving a licence to produce or sell products or services provides the whole programme of production and marketing for an interested farm, including the brand, logo and management methods. Marketing activity is agreed upon in the licence contract. Around the world there are many successful franchising networks. The idea of such a network is to ensure that the tourist will receive the same products and services, the same quality and the same price level in each facility of a particular franchising network. Clients recognize particular franchising networks and are willing to use their services. This system operates very well in catering firms such as McDonald's and Kentucky Fried Chicken and in hotel enterprises – Novotel, Sun Inn, Holiday Inn, etc. In this respect an example could also be the international chain Maize (introduced in Utah in 1996, Maize developed to be one of the largest enterprises offering maize mazes worldwide; in 2004, after 8 years, there were 460 mazes in five countries). It may be assumed that franchising networks will develop in agritourism on a large scale.

Agritourist Associations

An association is usually a non-profit organization, realizing statutory aims established by the founders of the association, with the organizational structure and mode of operation being defined in the statute passed by the founders and frequently verified by a respective court. Agritourist associations operate for the benefit of farmers engaged in agritourism. An association may serve several functions, e.g. advertising on the Net, through leaflets, etc. An association may offer a chance to cooperate, receive training and obtain information; associations organize conferences or even conduct joint research. The number of agritourist

Fig. 9.4. Logos of some European agritourist associations (from EuroGites, accessed April 2008).

associations is high. Usually in one country there are numerous local or regional associations. These associations join to form national agritourist associations. Finally there is a trend to form international associations. A large number of agritourist associations have been organized by farmers in the USA, in such states as Delaware, Hawaii, Massachusetts, Michigan, etc. Regional associations have also been founded, e.g. the Dude Ranchers' Association of America. Agritourist associations in Poland have an interesting history.

Agritourist associations operate in many European countries, e.g. Farm Stay in the UK, Asetur in Spain, Gites de France in France and Urlaub am Bauernhof in Austria (Fig. 9.4). The European Federation for Farm and Village Tourism 'Eurogites' comprises national associations from 24 European countries. This Federation claims that rural tourism is still a sleeping giant. The Federation initiated work on the development of standards. At present, the definition of 'rural accommodation' has been established, together with standards concerning quality assessment for rural accommodation.

In the years 1995–1997 the Tourist Industry Development Programme TOURIN II, financed by the European Union, was introduced. Within the programme the project entitled 'The Development of Tourism in Rural and Forested Areas' was realized. As a result, in 1996 the Polish Federation of Rural Tourism 'Hospitable Farms' was established (Fig. 9.5). The federation is a legal entity and, in 1996, 27 associations joined it. At present, it is a nationwide non-profit organization that unites 56 professional local and regional associations interested in the development of tourism in rural areas.

Agritourism and the Food Industry

As it turns out, agritourism is an activity that is available not only to farms. It is also pursued by enterprises that are completely unrelated to agriculture, but which count on high returns from the capital invested in food-processing activities. There are numerous examples. Small companies of the agricultural and

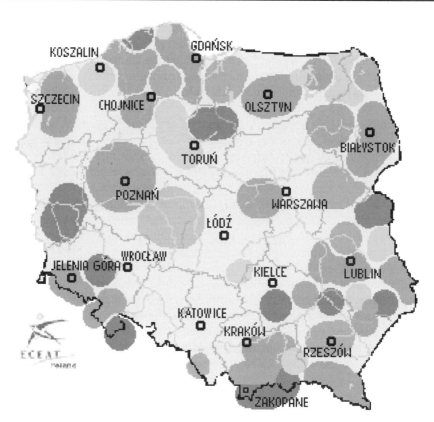

Fig. 9.5. The distribution and territorial range of regional agritourist associations belonging to the Polish Federation of Rural Tourism 'Hospitable Farms' (from Polish Federation of Rural Tourism 'Hospitable Farms', accessed April 2008).

food industry frequently use agritourism as the main market channel for their products, big companies use it as a tool for promoting their products whereas medium companies use it both as a source of considerable income and as a tool for promoting their products. Food industry enterprises enable tourists to observe the production process or they organize special outlets promoting or selling products. The element that tops the tour around the enterprise is a shop or company restaurant, where one can buy or consume the products whose production process was previously observed.

The observation of the production process is possible due to special paths and observation points for visitors. They are included in the architecture of the enterprise. The paths are separated from the production houses by glass partition walls and often have the character of galleries. Big enterprises of the food industry hope that showing the production process to tourists is a perfect method of promotion and even an effective investment in consumers' loyalty. Small-scale production enterprises expect tourists to purchase all the production at a price attractive to the enterprise.

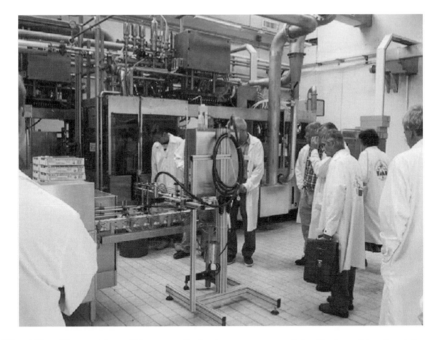

Fig. 9.6.   Observation of the manufacturing process: a dairy that enables tourists to observe butter production, Dairy Cooperative in Gostyń, Poland (photo by L. Przezbórska).

Many branches of the food industry use agritourism in the manner described above, especially enterprises manufacturing milk (Fig. 9.6), wine and beer. Meat plants also use agritourism for promotion, though the slaughter process and meat processing are not usually included in the sightseeing tour. The use of agritourism for the promotion of food products is more widely discussed in Part III of the textbook, which describes agritourism around the world.

Case Studies

1.   Investigate whether and to what extent there is any cooperation of agritourist farms in your region.
2.   What are the legal principles regulating the formation of an agritourist association in your region?
3.   What statutory tasks could be fulfilled by regional agritourist associations?

10 Agritourist Services and Products

Services, Products, Imponderables and Agritourist Attractions

Each business activity is aimed at providing services or offering products. The terms 'service' and 'product' are widely used in economics. A service has a non-material form and it could be, for example, mowing the neighbour's meadow, transporting the materials ordered by another farmer, etc. Hotels provide bed services and industry provides such products as cars, TVs, irons, food processors or clothes. A product is a material result of manufacturing or processing. In a farm the products are, for example, milk, beef animals, pork animals, cereal grains, etc. Sometimes the term product is also extended to services. In this approach agritourist services are also treated as products. The goal of agritourist activity is to provide services and sell tourist products.

In addition, agritourism is related to 'imponderables', i.e. unmeasurable values, which in particular refer to agritourist space, such as the beauty of the landscape, fresh air or rural architecture. Imponderables, though not a service or product, definitely facilitate the marketing process in agritourism. Agritourism also satisfies human emotional, aesthetic and social needs, especially the longing for rural serenity, fresh air and landscape undisturbed by architecture. Also agritourism satisfies the human's emotional need for contact with animals. Children especially feel very strongly the need for direct contact with animals. Even the most interesting nature films do not satisfy this need.

Many city dwellers long to stay in a traditional community, where the family, neighbourly or religious ties have not been broken, as is frequently the case in the city. There is also a longing for a lasting multigenerational traditional family preserving religious or other cultural values. Staying in rural areas satisfies the need for contact with fresh and healthy farming products and food produced according to home methods. Many people also feel the need to participate in the course of the production process on a farm.

Apart from the terms 'product', 'service' or 'imponderables' the term 'attraction' is also used in agritourism. Tourist attractions could be products, services, imponderables or a combination of these. An attraction means something specific that attracts the tourist to visit a particular place. It is unjustified to directly identify attraction with a product or service as some agritourist products and services are counted as attractions while others are not.

The Classification of Agritourist Products and Services

Modern agritourism offers an enormous range of products and services. Hence the need to classify them. Two main criteria for classification of agritourist products and services can be distinguished. The first criterion is related to their seasonal availability, the other to the division of agritourism into branches.

With regard to the seasonal character of their availability, agritourist products, services and imponderables can be divided into two groups, i.e. those that are constantly available all year round and those that are seasonally available. A large number of agritourist products and services have a seasonal character. Among them temporary services, products and imponderables constitute a considerable part. They are available at a given moment or they are not available. An example of temporary imponderables could be flowers blooming for a short period of time. A Carpathian meadow on which crocuses bloom for a few days is a perfect example of this.

Products, services and imponderables that have a seasonal character are usually more expensive than those provided all year round. Those services and products that are constantly available can be cheap. In many parts of the world it is possible to provide tourist services all year round. A holiday on Mediterranean islands can be relatively cheap because the high season there lasts almost all the year. In favourable conditions tourism is organized like an assembly line: some people leave and others arrive, every day excursions are organized and the planes, hotels and restaurants are fully booked. Thus all days in a year are used. In spite of this, in some countries where agritourist activity has a seasonal character, it is also treated as a cheap form of holiday. This results from the fact that agritourism is in the initial stage of development and farmers offer reduced or dumping prices (i.e. prices below one's own costs) to attract clients. This situation cannot last and pricing practices will slowly evolve. In the future agritourist products and services will be relatively expensive unless the agritourist process can be organized similarly to mass tourism. However, the prerequisite is the all-year availability of the products and services and a larger scale of provided services. Unfortunately, the individual character of agritourism may then be lost.

Figure 10.1 shows the division of agritourist products and services into different agritourist activities. Nine categories are distinguished: Agri-accommodation, agri-food and beverages, primary agritourism, direct sales, agri-recreation, agri-sport, agritainment, agri-therapy and cultural tourism. Within each specific activity category, products and services are distinguished. The following sections provide short descriptions of individual agritourist activities.

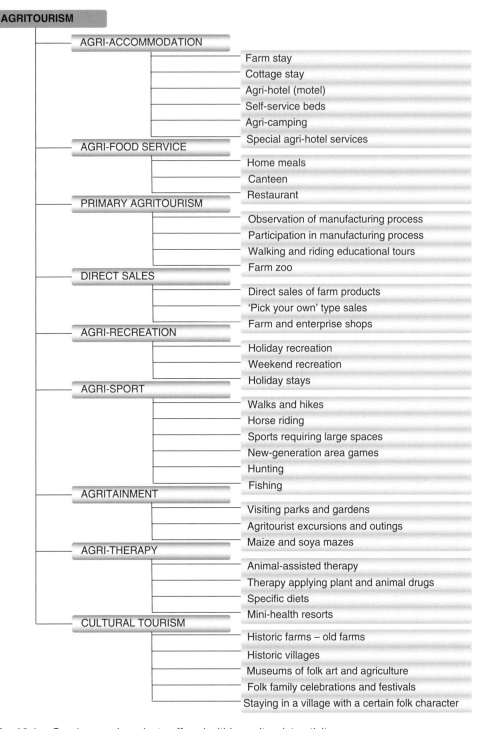

Fig. 10.1. Services and products offered within agritourist activity.

Popularity of Selected Agritourist Products and Services

Table 10.1 reports the popularity of selected agritourism products offered in 200 random farms that offer their services on the Internet (50 each from Great Britain, Italy, Poland and the USA). The analysis shows the agritourism products and services that are most frequently offered by agritourist farms in the selected countries. The most frequently offered agritourist products are farm stay, walks and hikes, visiting parks and gardens and different kinds of museums. The most infrequent products and services are specialist agri-hotel services, 'pick-your-own' type sales, new-generation area games, labyrinths and hippotherapy.

The products and services most linked to the production process are in general the most common products and services offered by agritourism farms. The reason for this is that offering these agritourism products does not require significant additional inputs as is the case with farm stays, observation of the manufacturing process, walks and hikes.

Agritourism farms in Great Britain and the USA offer a wider spectrum of products and services. Very carefully prepared offerings are characteristic for these countries so that even minor activities, such as wildlife watching and ballad presentations, are treated as separate products. The advantage of carefully prepared offerings is the potential to expand the number of potential clients interested in the enterprise and the ability to include even the most fastidious clients. In general, Italian agritourism presents a different style of offerings. Products and services in Italy are characterized by a very high standard and a weak linkage to agriculture. This means that it is mostly regular tourism in the rural areas.

The characteristic feature for Polish agritourism farms is the low level of specialization and in comparison with other countries agritourism is characterized by a rather small number of farms offering self-service beds. Offering that kind of service requires the creation of an appropriate infrastructure, namely separate buildings and individual lodgings that are fully equipped and of a high standard. Many farmers in the agritourism business do not hold sufficient funds for modernizing their farms in an appropriate way for agritourism.

Agri-accommodation Services and Products

The agri-accommodation industry comprises accommodation services provided by agritourist farms. The main goal of the hotel industry is to offer tourists accommodation for at least one night. Hotel services are offered by hotels and hotel networks of various types, with which the farmers providing hotel services are virtually not able to compete. This results from the fact that hotels cater for a different segment of clients from the agri-hotel industry. In particular, people looking for a bed in the city are not willing to take the offer of agri-hotels. A tourist staying or relaxing in a rural area needs a bed. The agri-hotel industry satisfies this need. It may be the exclusive service of an agritourist farm or an accompanying service. Despite this fact, competition or even tension is observed between hotels and agritourist farms offering accommodation.

Table 10.1. Popularity of selected agritourist products in Poland, Italy, Great Britain and USA (frequency %). (From Majkowski, 2004.)

		Country			
Category	Product	Great Britain	Italy	Poland	USA
Agri-	Farm stay	80	18	56	78
accommodation	Cottage stay	62	98	42	22
industry	Agri-hotel (motel).	16	32	16	28
	Self-service beds	96	88	38	30
	Agri-camping	6	0	36	12
	Special agri-hotel services	0	0	4	0
Agri-food and	Home meals	24	12	80	28
beverages	Canteen	0	2	4	8
	Restaurant	6	8	4	10
Primary	Observation of manufacturing process	46	12	38	40
agritourism	Participation in manufacturing process	14	8	32	32
	Walking and riding educational tours	34	2	10	22
	Farm zoo	20	6	30	22
Direct sales	Direct sales of farm products	4	34	20	22
	'Pick your own' type sales	2	6	6	14
	Farm and enterprise shops	14	0	4	12
Agri-sport	Walks and hikes	94	96	98	96
	Horse riding	10	20	34	38
	Sports requiring large spaces	44	74	86	76
	New-generation area games	2	0	2	0
	Hunting	4	0	8	14
	Fishing	44	6	46	42
Agritainment	Visiting parks and gardens	82	88	64	82
	Agritourist excursions and outings	20	80	52	56
	Maize and soya mazes	0	0	0	2
Agri-therapy	Animal-assisted therapy	0	0	2	0
	Therapy applying plant and animal drugs	0	0	2	0
	Specific diets	30	4	8	8
	Mini-health resorts	2	2	4	10
Cultural tourism	Historic farms – old farms	26	40	0	26
	Historic villages	4	44	10	6
	Museums of folk art and agriculture	28	12	32	24
	Folk family celebrations and festivals	8	4	14	38
	Staying in a village with a certain folk character	0	0	0	2

In order to eliminate this misunderstanding, a solution is adopted of limiting the number of accommodation places in the agri-hotel farms. For example, in Italy on one farm there may be a maximum of 45 beds, while in Poland it is a maximum of five rooms for rent as accommodation. Thus the agri-hotel industry differs from the hotel industry, which is expressed by the specific services that are offered. The following categories of agri-hotel services can be distinguished:

- Farm stay;
- Cottage stay;
- Agri-hotel (motel);
- Self-service beds;
- Agri-camping;
- Special agri-hotel services.

A farm stay is an accommodation service provided by an agritourist farm for tourists who want to stay there for at least one night. This form is very common, for example, in New Zealand (Fig. 10.2), which is visited by numerous tourists from Asia, Europe and America. Tourists rent cars or motorcycles and reach the most remote corners in the country. They often choose to stay on agritourist farms overnight. The pattern of service provision on such farms is very similar. Tourists need to book in advance, because farmers have to prepare to receive guests. For example, buying extra food takes time, given that the nearest shop is often 20–40 km away. Tourists usually arrive at the farm in the afternoon, before dinner. The hostess welcomes the guests, trying to make a nice and friendly atmosphere from the very beginning. The guests take a while to unpack their belongings and have a wash. Then they chat with the hosts and in the evening they have dinner, which is the main meal of the day. The host is responsible for chatting and is usually prepared to discuss a range of topics. About 10 pm the

Fig. 10.2. Farm stay in New Zealand, bed and breakfast, Kohinoor, Ohakune (photo by M. Sznajder).

hosts want to go to bed and they expect the same from their guests. The next day is a working day for the owners of the farm. The following morning the guests get breakfast. Usually they can choose between cooked breakfast and continental breakfast. Continental breakfast on New Zealand farms is smaller than cooked breakfast and usually includes muesli (flakes), fruit, yoghurt, milk and tea or coffee. If the tourists like, a tour of the farm is organized, which takes 2–4 hours. On return there is a guest book on the table, which must be signed and then a typical question is asked: tea or coffee? It is a sign that it is time to pay for the stay and leave the farm. Similar rules apply to staying in a cottage (rural tourism house). A tour around the farm is not offered here unless the owner of the house has signed a contract with the neighbouring farm.

Agri-camping is a camping service provided by agritourist entities. Usually it is possible to pitch a tent on a campsite and use the bathroom, shower and toilet. Self-service beds are an interesting form of the hotel industry. The owner offers all the necessary hotel services, but in the evening there is nobody awaiting and welcoming the arriving guests. The tourists are informed about the room access code or the place where the key can be found, which is enough to use the service.

The agri-hotel industry may also offer alternative services (Fig. 10.3), such as sleeping on hay. Alternative services cannot be a mass service, though they are often combined with survival camps, which are currently in fashion in some places. However, such services involve certain risks for the farm, for example, fire hazard, and for the tourists themselves – low sanitation standard, skin diseases hazard, phosgene poisoning hazard, etc.

Fig. 10.3.   Alternative bed service. A wooden hut (photo by Ł. Sznajder).

It must be stressed that agri-hotel services not only need to comply with the requirements specified by law; they may also need to be evaluated by auditors. Auditing organizations give quality signs depending on the standard of the services provided. In order to receive such certification the farmer has to pay a certain auditing fee. The problems of service quality in agritourism and its control are discussed in Chapter 9.

Agri-food Services and Products

Agri-food services are an integral part of agritourism. Tourists staying temporarily or seasonally on a farm need to have food. They can use a restaurant, eat their own meals or use the food services offered by agritourist farms. Usually four forms of agri-food services can be found:

- Home meals eaten by tourists together with the farmer's family.
- Picnic meals eaten in the fields.
- Canteens, i.e. separate places for tourists, who eat meals without the farmer's family but according to the menu and timetable imposed by the farmer.
- Restaurants run by farmers or food-processing enterprises, which provide meals not only for holidaymakers staying in a specific farm but also for outside guests.

The diversification of agri-food services may be enormous. This may apply to the origin of products, the ingredients and the number of meals, as well as the time and place where the meals are served (Fig. 10.4). A standard service offered by agritourist farms is bed and breakfast, usually with a choice between cooked and continental breakfast.

Food items may partly be produced by the farm or they may all be purchased. By serving home-grown food, the farm increases its net income. As regards the time when meals are served, we can distinguish between breakfast, lunch, dinner, morning and afternoon tea (or coffee), tea and supper. As regards the place where meals are served, we can distinguish between meals served in the dining room (kitchen), outside the building or in special restaurants (in particular this applies to agritourist entities related to the food-processing industry). In respect of dining habits, we can distinguish between meals specific to a particular family, regional cuisine and national cuisine. We could also distinguish between fast and feast meals. The feature of agri-food and beverages is its diversity, which ensures that the meals served in one farm differ from those served in another. Competitiveness and tension are also found between traditional restaurants and agri-catering service providers. In order to eliminate these tensions, different legal solutions have been implemented. Most frequently they include limitations concerning the minimum amount of food that may come from one's own farm or from neighbouring farms (so that the food is produced locally) and the maximum amount of food that may be purchased. In Italy, for example, these proportions are 30% and 40%, respectively.

Fig. 10.4.   An example of a location for agri-food services: open-air, no roofing, on compacted straw (photo by M. Sznajder).

The Products and Services of Primary Agritourism

Primary agritourism comprises the categories of tourist products and services that are directly related to the production process on a farm. The products and services of primary agritourism include:

- Observation of the production of crops and livestock.
- Observation of food and beverage processing.
- Participation in the production of crops, animals, foods and beverages.
- Educational tours.
- Farm zoo, safari, direct contact with domestic animals or nature on the farm.

Observation of the production process

This enables tourists to observe the whole or the most interesting fragment of the production process. In crop production most often it is observation of only part of the process because the whole process lasts many months. The most interesting moments are sowing, blossoming (e.g. orchards, brassicas) and harvesting. Tourists can also familiarize themselves with systems of land utilization. A tourist-oriented farm should have a room with plans showing the development of systems of land utilization. Observation of the livestock-producing process is relatively easy in the case of livestock grazing in pastures but difficult or in fact impossible on most modern livestock farms in Europe.

The food-processing industry, especially small companies, more and more frequently try hard to attract tourists. There are special spaces arranged behind glass, which separate tourists from the food processing so that it can be undisturbed. Figure 9.6 shows an example of observation of the manufacturing process: the production of butter in a dairy.

Participation in the production process

The participation of tourists in the production process is an agritourist product that involves certain risks. Thus, in this case, it is necessary to have liability insurance in case of an accident. Under supervision, agritourists can participate in crop and livestock production and, much less frequently, in food processing. Guests can take part in feeding animals, milking cows, driving animals in pasture, etc. In crop production their participation may apply to sowing and harvesting. New technology in agriculture makes tourists' participation in the producing process difficult but does not exclude it.

Educational tours

Educational tours are becoming a more and more popular form of demonstration of not only farms, villages and forests but also national parks to tourists. Tourists can choose between a walking educational tour or a ride on a cart or mechanical vehicle. During the walk people have particularly close contact with nature. There are also farms that have educational tours allowing observation of the producing process. Farms of varied terrain configuration are particularly suitable for this purpose. In units whose area ranges between 30 and 50 ha, whole families can walk along educational paths observing animals, crops, birds and various stages of farming production. The idea of educational tours consists in the fact that tourists individually walk around a certain area of particular interest according to an arranged order. There are stops with benches on the way and information boards indicating what deserves special attention in a particular place.

Farm zoo, safari, direct contact with domestic animals

These are a special attraction for children. Some farmers adapt all the area of their farm to arrange a farm zoo or terrain for a bloodless safari. Apart from the possibility of admiring exotic animals kept in decent conditions, it is also possible to have direct unlimited contact with such animals as poultry, rabbits, goats, sheep or calves. However, it must be remembered that animals must be accustomed to contact with people and suitable for the age of the children. Children, in turn, must remain under adults' supervision so that they will not harm the animals. Figure 10.5 shows an example of a farm that enables direct contact between people and animals.

Fig. 10.5.   Direct contact with domestic animals: farm safari in Świerkocin, Poland (photo by L. Przezbórska).

Two animal feeding situations can be identified. The first situation refers to animals kept in farm zoos and the second refers to wild animals. Feeding of wild animals is often very controversial. In the first situation the owner of the animals offers visitors the possibility to buy food to feed one's animals. Feeding animals gives huge pleasure to visitors, especially children.

Feeding animals is controversial, however. People cannot resist the willingness to feed both wild and domestic animals. One can quite often observe city inhabitants and tourists feeding swans, ducks, pigeons and even squirrels. It happens that in cities some inhabitants earn money by selling feed for birds to tourists, as is the case in Kraków or Buenos Aires. This results in flocks of birds gathering in feeding places. However, the position of a good-natured man feeding 'poor' and 'hungry' animals is more and more frequently criticized. Today many people think animals should find food for themselves. Wild animals that are additionally fed by people become troublesome. Historic houses are polluted by bird excrement, ventilating ducts and gutters are blocked by birds' nests, animal diseases such as avian tuberculosis are transmitted. Small water reservoirs with food thrown into them do not make a positive aesthetic impression. Animals become more and more people-dependent and bolder and bolder. In the Tatra Mountains bears reach for tourists' rucksacks with food and come to dustbins near hostels, where they can find scraps. In Gibraltar monkeys snatch bags with food, which often also contain IDs and money. Because of their importunity swans become dangerous to children and kea parrots fed by tourists on the South Island of New Zealand have become so importunate that they can peck at rubber seals in cars left in car parks or do a lot of damage to huts or garages. In open-air restaurants birds may even take your food off your table. This causes a lot of problems to both tourists and residents. In many places feeding birds is totally prohibited and caretakers fine tourists for doing so. Therefore, feeding wild animals in public places cannot be treated as a tourist attraction. However, many agritourist farms include feeding animals, especially poultry, in their package of attractions.

Direct Sales

Due to the fast development of modern wholesale and retail networks, whose goal is to sell mass products to as large a number of consumers as possible, there is a danger that local products manufactured in small amounts will disappear completely, because there will be no effective markets for them. Agritourism provides a market opportunity for such commodities. It is not only agritourist farms that are interested in direct sales but also local rural communities and increasingly also enterprises manufacturing or processing food, especially small and medium-size enterprises.

Within the branch defined as direct sales we can list the following:

- Direct sales of farm products;
- 'Pick your own' sales;
- Stands with agricultural produce;
- Farm or enterprise shops.

Direct sales of farm products

A tourist observing or participating in the manufacturing process is ready to buy the products whose manufacturing he observed or in which he participated himself. Thus, the farmer offers tourists the possibility to stay on his farm and then buy the products manufactured on the farm. Usually this applies to the possibility of buying such fruit as strawberries, raspberries, currants, gooseberries, apples, pears, etc. Sometimes tourists can buy animal products. Direct sales are an important agritourist service. Figure 10.6 shows a farm in New Zealand where city people buy fruit and vegetables. Selling fruit, vegetables, blueberries and mushrooms by the road is an interesting custom in Poland and many other countries. There are stalls offering fruit and vegetables to tourists by some roads, especially those leading to Warsaw or Silesia.

Direct sales – 'pick your own' type

This is a specific form of agritourism. Farmers prepare plantations so that at harvest time tourists can pick fruit themselves. For this purpose an orchard or berry patch should have intercrops of low mown grass so that tourists will not dirty their clothes. The charge for the service is a problem as part of the fruit may be

Fig. 10.6. Direct sales of farm products – a farm in New Zealand (photo by L. Przezbórska).

eaten while being picked. Tourists may also pick the finest-looking fruit leaving a large amount of other fruit on the trees or shrubs.

Farm or enterprise shops

These are another agritourist option. Such shops may offer not only their own products but also the products made by the rural community. Many food-processing plants offer their own products for sale, treating it as a kind of promotion. In particular this applies to dairies, wineries and distilleries.

Stands with agricultural produce

These are an opportunity to increase income for farmers who sell their products to customers passing by in their cars. Stands are put along frequented roads.

Fig. 10.7.   Direct sales: stands with fruit, Ahmadabad, India (photo by M. Sznajder).

Stands may be found in many countries worldwide. Fresh fruit (Fig. 10.7) and vegetables and occasionally eggs are sold in this way.

The Products and Services of Agri-recreation

Agri-recreation is another branch of agritourism. It is considered to be the major reason for agritourism, at least in contemporary Poland. Agri-recreation combines the hotel and food services for a longer period of time. The idea of agri-recreation boils down to spending free time away from home on a farm or in rural accommodation. The products and services of agri-recreation are classified according to the length and season of stay in the following way:

- Holiday on a farm offered by vacation farms;
- Short weekend holiday (1–3 days) or longer holiday (e.g. during long weekends);
- Staying at Christmas, Easter or New Year.

A vacation farm is an agritourist farm receiving holidaymakers for a holiday lasting from 1 to 4 weeks. In Poland the tradition of a 2-week holiday is common. Vacation farms usually receive holidaymakers from June to the end of August. This service is the dominant form of agritourist services in Poland.

The Products and Services of Agri-sport

Sport is a branch of human activity that is not usually associated with agritourism. However, agritourist farms have particularly suitable conditions to offer the possibility to do some types of sport. These predispositions are particularly visible when a sport involves keeping animals or if a large and diversified space is necessary to do the sport.

Within agri-sport the following categories can be distinguished:

- Walks and hikes;
- Horse riding;
- Sports requiring a large space;
- New-generation area games;
- Hunting;
- Fishing.

All types of horse riding are usually listed as a standard agritourist sport. However, within one group many different services can be offered: controlled riding on a horse led with a bridle or lunging rein, horse riding lessons, rides on a wagon or cart. For this reason one farm offering horse riding is not similar to another.

Agri-sport may also include all types of walks and cycling. Farms can offer all sports requiring large space, for example, golf courses. They can also include in their offering various types of conventional and extreme sports. Therefore, the use of the term agri-sport in reference to sports by agritourist farms seems to be justified. Some farms located in suburban areas build and maintain tennis courts or even use large areas for golf courses in order to increase their profitability. However, it must be remembered that the preparation of a high-standard sports facility by a farm requires high investment.

Extreme sports, for example for BMX bikes or paintball, are an interesting development. These sports require significant investment and a fairly diverse area. A farm may offer the rental of sports equipment, e.g. balls, canoes or bicycles.

What can also be included in agri-sport are new-generation area games, which have recently become fashionable. New-generation area games are also features of other games. Live-action role play (LARP) is an example. Each player has a role to play. The characters live in a world defined by the convention of fantasy. Many characters in area feature games have been borrowed from the classic writer of the genre, J.R.R. Tolkien. The first LARP in Poland, Morkon, is a feature game played in the forest and hills area (Fig. 10.8). The game is composed of a series of consecutive editions, e.g. The Dark Wave, The War of Gods, The Child of the Dragon, etc.

In addition to these sports, hunting and fishing (Fig. 10.9) must be included in the agri-sport category as rural areas are the primary places where people can participate in these activities. Within hunting we may distinguish bloodless safari, falconry and traditional hunting. Hunting and bloodless safari are not typical agritourist products, although they are closely connected with farms. Generally falconry, being much less common, may be considered to be an agritourist activity. It is crucial to respect legal regulations concerning these services.

Fig. 10.8. New-generation games – Morkon, Poland (photo by Ł. Sznajder).

Fig. 10.9. Agri-sports: fishing from a farmer's boat on a lake in Poland (photo by M. Sznajder).

Agritainment

People today have many opportunities to experience attractive entertainment. In fact, it is hard to define what entertainment is and what it is not. Both tourists visiting big cities and rural areas and their inhabitants can make use of a wide range of entertainment. Some entertainment can be classified as the so-called agritainment. The use of the term agritainment is justified by the fact that such entertainment is only available on farms or in rural areas or possibly in green areas in cities.

Three categories may be included in the agritainment group, namely:

- Visiting parks and gardens;
- Agritourist excursions and outings;
- Maize and soya mazes.

Historically the oldest form of agritainment, which appeared long before the concept of agritourism originated, was staying in and visiting city parks and gardens. They can be found almost everywhere around the world. Local communities do their best to have a nice-looking park or garden. Various events are organized in such places, e.g. concerts and festivals. Visiting parks and gardens is the favourite form of entertainment of city inhabitants.

Excursions and outings to rural areas, including picnics of various types, are another form of agritainment. Such events are usually organized by travel agencies, which have to collaborate with the owners of lands or farms on which the events are to take place. Agritourist excursions may take various forms. It could be an ordinary trip, but also a study excursion, an agritourist adventure (i.e. a trip, often unorganized, full of impressions of risk) or finally an exploration trip. An agritourist trip may be organized as a form of recreation but also as an educational trip. The trip is organized for a group of children or adults, usually city inhabitants. It lasts from one to several days and usually it involves frequent changes of the place of staying in accordance with the planned route. An agritourist study trip is organized for specialists who want to learn about the food economy sector of a particular region or country. Study trips usually include visiting farms on the route but also food industry enterprises and companies trading in food items. An escapade is in fact a form of travel for older youth. Usually it is not well organized. An exploration trip applies to a small group of people whose goal is to reach the places that are not visited by tourists, learning about the natural and social characteristics of the area, including the agriculture.

A specific and well-developing service is connected with school trips. Lessons in the natural sciences, biology, geography or history are excellently related to practical visits to farms. Farms serving school trip traffic and so-called green schools are being established worldwide. Since this segment of target customers has a lot of potential for development, an increasing number of agritourist farms focus on this type of services.

In recent years a new generation of agritainment has developed rapidly. It requires farmers to devote a considerable area of their farm to the activity. Investment funds are also necessary. The maize maze shown in Fig. 10.10, pizza farms

Fig. 10.10. A view from a plane of a maize maze belonging to the American Maze Company: the Aztec Calendar, Pacific Earth Farm in Camarillo, California (from Don Frantz/American Maze Company, reproduced with permission).

and field games in the area with constructed straw structures are examples of this type of entertainment.

Maize mazes are built on farm fields usually near big towns so as to guarantee a constant inflow of people interested in this type of entertainment. Mazes seem to herald the appearance of new generation of agritainment. This form of service has become particularly common in the United States. There has also been established an international company, The Maze, which propagates the idea around the world. Mazes can be found in the United States, Canada, France, New Zealand and Poland. The objective of the game is to reach the destination, which is 'Geronimo's eye'. There the winner gets the prize. To make the search easier one can use a special map. The shortest route to the eye of the maze is about 800 metres long. For little children there are mini-mazes arranged in soya fields (Mayse, 2003).

Structures made out of straw are often used for agritourist purposes. Some are built using a wooden frame and others are made using bales of compressed hay or straw. Sometimes play sites and fortifications are developed. Sometimes they are designed to be models of historic villages. At other times, they are designed to be suitable venues for new-generation games.

Agri-therapy

Therapeutic services may also be related to agritourism. In fact, it is also a developing branch of agritourism. The name agri-therapy is justified by the fact that the therapy must take place on a farm or in a rural area. Currently agritourism offers at least four products and services of agri-therapy, namely:

- Animal-assisted therapy.
- Therapy applying plant and animal specifics:
  - Aromatherapy;
  - Apitherapy.
- Specific diets.
- Mini-health resorts.

People may benefit from animals in various ways, including the comfort of physical contact with them, reducing loneliness and increased opportunities for meeting other people via the pets. Different kinds of animals are used in therapy, including pets such as dogs, cats and birds, but also horses, elephants, dolphins, rabbits, lizards and other small animals. The Delta Society, an international, non-profit, human service organization providing and promoting human health and well-being through interactions with companion animals, has distinguished between two kinds of contacts with animals: animal-assisted activities (AAA) and animal-assisted therapy (AAT) (Delta Society, accessed April 2008).

Animal-assisted activities provide opportunities for various benefits from direct contact with animals: motivational, educational and recreational benefits to enhance the quality of life. The main goal of animal-assisted therapy is a directed intervention in which a specifically selected animal is an integral part of the treatment process. Animal-assisted therapy is aimed at improving the physical, social, emotional and/or cognitive functioning of a patient, but also providing educational and motivational effectiveness for participants. AAT is designed to promote improvement in human physical, social, emotional and/or cognitive functioning. It can be provided individually or in a group. During AAT, therapists document records and evaluate the participant's progress.

Many different animals can be utilized to provide AAT. Therapeutic riding programmes improve the motor skills and coordination of the physically challenged (hippotherapy or horse therapy). Pets help inmates in correctional facilities and juvenile offenders to learn empathy and compassion.

Hippotherapy is a method of rehabilitation for disabled and handicapped people after physical and mental diseases. The presence of a horse makes the therapeutic method exceptional and unique. However, it is closely related to other rehabilitation and therapeutic methods. There are three categories distinguished: physiotherapy on horseback, psycho-pedagogical horse riding and therapy with a horse. The aim of hippotherapy is to stimulate recovery as well as physical fitness and mental capacity by means of a horse and horse riding. By this means balance disorder is reduced, defence reactions improve and the locomotive capacity increases. Hippotherapy provides contact with the animal and nature, stimulates psychomotor development and improves eye and motor coordination and orientation in space. It is specified as hippotherapy because of the horse, which participates in the therapy. Hippotherapy enables increasing the concentration capacity and maintaining organized activity, increases motivation to exercise, develops independence, increases self-esteem, provides relaxation and reduces neurotic reactions.

In many countries dolphins are commonly used in different kinds of thera-pies, for example, when autistic children swim with dolphins. There is a special dolphin-assisted therapy (DAT) programme, which is a result of more than 30 years of continuous scientific work with dolphins in the USA, Australia, Turkey and Ukraine (Dolphin-Assisted Therapy, accessed April 2008). According to dif-ferent sources, dogs (dog therapy) have been used in health facilities such as nursing homes (e.g. activities of the Delta Society). In recent years, therapy dogs have been used to help children overcome speech and emotional disorders. It is important that therapy dogs be of all sizes and all breeds and colours. The most important feature is the temperament of a dog. A good therapy dog must be friendly, patient, confident, at ease in all situations and gentle. Therapy dogs must enjoy human contact and be content to be petted and handled, sometimes clumsily. However, cats (feline therapy) and their owners are becoming more and more involved in therapeutic interaction. As pets are an integral part of agri-tourism farms, they can easily be used as a part of AAA or in AAT when a profes-sional therapist is available (Pets and People's Feline AAT Program, accessed April 2008).

Agriculture also offers therapeutic specifics, which may be applied only at the place where they are manufactured, i.e. on the farm. This applies to, for example, aromatherapy (i.e. therapy using aromas, a method of treatment dat-ing as far back as the history of civilization; it is based on the use of advanta-geous properties of essential oils contained in plants), apitherapy (i.e. a method of treatment, prevention and promotion of health using bee products and prepa-rations, so-called apitherapeutics, made using these products) or kumis therapy (i.e. the application of a fermented dairy product traditionally made from mares' milk). It should be expected that in the future the number of specifics offered by agritourist farms will rise constantly.

A tourist arriving at an agritourist farm can also count on the possibility of a specific diet to which curative, slimming and other properties are ascribed. In some countries within the area of some farms small springs of curative water or hot springs can be found. Such farms can organize mini-health resorts.

Cultural Tourism

Agritourism may also offer products and services related to cultural tourism. These services may be offered by individual farms as well as whole local com-munities. Rural communities frequently manufacture specific cultural products that distinguish them from other groups of people. Therefore ethnographic prod-ucts and services are usually offered to tourists by whole rural communities rather than an individual farm. The ethnographic products and services offered by the rural community include:

- Historic farms – old farms;
- Historic villages;
- Museums of folk art and agriculture;
- Folk family celebrations and festivals;

• Staying in a village characterized by specific folk character (dialect, art, attire, buildings, etc.).

Farms constantly change their manufacturing technologies. This means that modern technology is completely different from that in use several years ago. Harvest is an example. In the past corn was cut with a sickle and then bound into sheaves, which were transported to the barn. In autumn or winter corn was threshed and finally ground by means of a quern. Another important machine powered by tractive force was the treadmill, which is no longer known today. The sickle was superseded by the scythe, the scythe by the mower, the mower by the sheaf-binder and the sheaf-binder by the combine harvester. The threshing machine superseded flails, the combine harvester superseded the threshing machine and treadmills became useless. Agritourist farms usually present modern manufacturing technology. However, historic farms conducting the manufacturing process according to the technology of a particular century can be organized. Managing a historic agritourist farm requires a high investment of labour and it is very costly. That is why this agritourist service is rarely offered. Instead ethnographic parks and museums are established where specimens, posters, etc. presenting old manufacturing technologies are collected. In some regions of the world real-life examples of traditional production have been preserved; for example, the farms run by members of the Amish religious group, who reject many contemporary technological inventions. The Amish are a very conservative Christian religious group. The members speak only the German dialect called Pennsylvania Dutch. At a maximum they complete eight years of schooling. They do not use cars, electricity, radio or television. Marriages to people from other religious groups are forbidden. They are characterized by special attire. They do not take photographs. In spite of the fact that the community tries to remain isolated from the secularized world, the commercialization of their culture has become their main source of income. Tourism has the greatest influence on commercialization. Lancaster County, Pennsylvania, is their centre.

In order to commemorate the history of farm or village development, local communities or even individuals arrange historic villages. There are two types of historic farms. The first type is a natural village, located where it was founded; sometimes it is still inhabited (Fig. 10.11) or it is deserted by the inhabitants on purpose and conserved for visitors. The other type is a cultural park (or ethnic village), whose houses and structures have ethnographic value and are deliberately assembled in a particular area. There are various types of parks, museums, living cultural museums and museums of agriculture (Fig. 10.12). In living cultural museums people play the same roles as they did centuries ago. Maintaining a living cultural museum is extremely expensive and it requires a high inflow of tourists.

The Portfolio of Agritourist Products and Services

The above review of products, services and imponderables indicates that agritourism can offer a wide variety of services to clients. An agritourist farm may offer

Fig. 10.11. Historic towns and villages in New Zealand on North Island: the town of Whangamomona by road no. 43, gradually deserted by the inhabitants and transformed into a historic town. The hotel still receives guests (photo by M. Sznajder).

Fig. 10.12. Historic towns and villages: country museum in Kluki, Pomerania region, Poland (photo by L. Przezbórska).

one agritourist product or service, but most frequently it is a range of various products and services. Such a set is called the portfolio of agritourist products. The farm has a chance to achieve good results if it is able to organize a portfolio that is both attractive to tourists but also different from the portfolio of products and services of other farms.

Case Studies

1. List products and services offered by agritourist farms in your region. Compare this list with the list of agritourist products in Fig. 10.1. What are the differences?

2. Select any product or service from your list and prepare details concerning this product so that it may be offered to tourists.

3. Decide how important values may be included in the product to make the offered product or service more attractive.

11 Agritourists

Tourist and Income Flows

Tourists are an indispensable element of the functioning of agritourist farms and it is thanks to them that financing the activity of these farms is possible. The inflow of tourists will be referred to as tourist flow and the inflow of agritourism income. The numbers of tourists using agritourism are surprisingly high. USDA estimated that over 62 million people aged over 16 visited agritourist farms within 1 year (Wilson *et al.*, 2006). This does not include children aged below 16 years, estimated at 20 million. Data given by the Travel Industry Association of America suggest even 87 million. In the literature on the subject there are different amounts quoted concerning the tourist flow to the agritourist farms. For example, within 5 weeks the farm belonging to W.R. Crows, situated near Tolleson, Nebraska, received as many as 100 thousand people, who took part in the grape and pumpkin harvesting and travelled around the farm on hay carts (Good to Grow Farmers Enhance Their Prosperity with Agritourism, 1997). Another farm, Prairie Rose Chuckwagon Super, situated in Boston, 15 miles away from Wichita, Kansas, serves more than 60 thousand clients a year (Dinell, 2003). Barnyard Buddies Farm in East Plano, Texas, is visited by 40 adults with children every day, which amounts to 30 thousand people a year (Mayse, 2003).

If the tourist flow exceeds 20 thousand people a year, we can regard it as considerable. The literature usually quotes the tourist flow amounts ranging from 7 thousand to 15 thousand a year. The tourist flow depends on a number of factors: the vicinity of a city, good transport, products and services offered by the farm, advertising, etc. In addition, the flow changes depending on the time of the day, the day of the week or the season of the year. What results from observation is the fact that the highest tourist flow is early in the afternoon, Friday to Sunday and during the holidays, whereas the lowest flow is in the morning, on Mondays and in late autumn and winter.

The income flow is the function of the tourist flow and the pricing strategy as well as willingness to pay for these services on the part of tourists. The flow of agritourism-related incomes is generated from different sources. One of them is the admission charge. Studies showed that 5% of agritourists want to visit agri-tourist farms free of charge, 34% are willing to pay US$1–5, 34% US6$–15 and 5% over US$15. A total of 16% respondents claimed that price does not have an effect on their decision to visit an entity. The same studies indicated that 1% of visitors spend less than US$5 during the visit, 17.1% an average of US$10, 24.4% an average of US$20, 19.5% an average of US$33 and 15.9% more than US$40. If a statistical tourist leaves only US$20 on the farm and if 7000 tourists visit the farm, the yearly income is US$140 thousand, which is a considerable income in itself. Therefore the knowledge of clients and their agritourist behaviour is extremely important for people running agritourism businesses. For this purpose, they usually conduct statistical research concerning segmentation of customers.

The Segmentation Criteria

This section refers to the consumers of agritourist products and services as agri-tourists. A person offering agritourist products should know where to look for potential clients and learn what their needs, tastes and expectations are. Agri-tourists, like any other group of consumers, differ in some tastes, on the one hand, but also have similar tastes, on the other. A common characteristic of agri-tourists is their willingness to stay on a farm, observe production processes, have contact with animals, etc.

However, not all tourists are interested in agritourism. Each person is a unique individual and it rarely happens that people have identical tastes, wear identical clothes and think the same. In spite of this we can group all potential consumers according to certain typical features. Such grouping is called segmentation. The segment is a group of consumers, here agritourists, characterized by common features. The segmentation of consumers is of fundamental importance in market analysis, especially in marketing activity. Products and services are prepared for a specific segment of clients, e.g. hippotherapy is addressed to people suffering from motor and emotional disorders, direct contact with animals is offered especially to little children. It is important that the segment should be big enough to finance the offered product or service.

There is no universal method of segmentation to be applied to the tourist market; thus many different criteria are used, depending on needs and conditions: nationality, style of tourism, manner of travel and lifestyle but also demographic, economic and psychological characteristics.

According to Middleton (1996), in tourism marketing four groups of segmentation criteria may be distinguished: descriptive variables (i.e. descriptive market segmentation), explanatory variables, variables characterizing lifestyle and geo-demographic variables.

The first group, i.e. descriptive variables, is widely used in tourism due to its simplicity, clarity and easy statistical analysis. Using descriptive variables we may survey the market using demographic characteristics (age, sex, income per capita, stage of family development, etc.), geographical criteria or the criterion of response to the product.

When using explanatory variables we may segment the market on the basis of such characteristics as reasons to travel (visiting relatives, business trips, improvement of health, visiting historic monuments, sports, etc.), benefits connected with satisfying needs (i.e. benefit segmentation), psychographic factors (adaptation of products and service packages to the traits of individuals from specific cultural and civilization regions) and social roles of reference groups (both formal and informal).

Segmentation according to variables characterizing lifestyle (i.e. way of life) or cultural affiliation refers to attitudes, behaviour of individuals defined by their system of values, belonging to specific social groups, religion, political opinions, etc. According to Altkorn (2000), the idea of segmentation by lifestyles may be very well characterized by names of consumer groups: yuppies (young, upwardly mobile professionals), dinkies (double incomes, no kids), wooppies (well-off old people) and glammies (the greying, leisured, affluent, middle-aged sector of the market). In turn, the division according to geo-demographic factors (geo-demographic market segmentation) is based on the criterion of the place of residence and living conditions and is 'a certain identification of the social position' (Altkorn, 2000).

Modern agritourism, however, requires more advanced segmentation of tourists. The basic and most frequently applied criteria for segmentation of consumers of agritourist services are demographic, economic and psychosocial characteristics.

Demography is the science that examines and describes the population of the world, a specific country or a region. Most population characteristics are important to agritourism. In this chapter we shall characterize the population by overall population size and distribution of the population according to age, sex, place of residence, nationality, religion, education and size of household.

The second criterion for segmentation is economic characteristics, e.g. disposable income per person and sources of income. The third criterion is the behaviour of agritourists. This behaviour includes many elements, e.g. time for recreation, preferred places for recreation, manner of travel to the agritourist farm and preferences concerning products and services.

Lifestyle segmentation of clients is most frequently used by manufacturers and service providers. Some people like to get up early in the morning while others get up very late; some prefer the calm countryside, others the noisy city. The fourth criterion for segmentation is the values that influence the people participating in tourist activity, including their preferences, values and tastes. Lifestyle segmentation boils down to distinguishing relatively homogeneous groups of society in terms of their aspirations and behaviour. As long as agritourism intends to attract an increasing number of clients, there is value in developing lifestyle segmentation skills. Statistical examples are given for the USA and Germany, i.e. countries with well-developed agritourism, and for Poland, with its developing agritourism.

Population

The knowledge of demographic trends, especially according to age groups within a population, is extremely important to agritourism. The following division is helpful: younger children, older children, youth, fully active adults, elderly people.

Little children up to the age of 5 take a keen interest in animals; however, they need to be constantly attended by adults when contacting animals. Young children aged 5–7 years can be very grateful consumers of agritourist services if the products and services offered are safe and satisfy their emotional needs. Their emotional needs are primarily satisfied by direct contact with docile, especially young, animals. However, children must be under control in order not to harm the animals. Older children aged 8–14 years prefer physical activity as in an adventure playground or obstacle course. They can interact with larger and older animals. They need a more energetic agritourist experience than younger children.

Youth usually expect to have contact with their peers and they are usually mobile. An agritourist service for this group should allow for the fact that young people want to stay together and give them an opportunity for competition and a chance to show off. People in their prime often want to avail themselves of short, weekend experiences. Older people above all expect to find peace and meet their peers. Therefore, different types of agritourist products and services should be addressed to each of the groups listed.

The distribution of populations in terms of age groups and presented graphically is called an age pyramid. There are different types of age pyramids. Figure 11.1 presents the age pyramid for the world population. Young people still predominate worldwide, although it is no longer the rule.

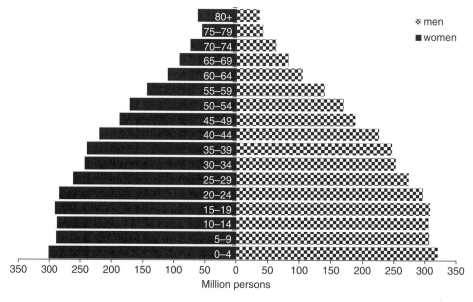

Fig. 11.1. The age pyramid of the world population in 2007 (from US Census Bureau Current Population Survey, 2007).

The age pyramid for wealthy countries looks very different. Figure 11.2 presents an example of the age pyramid for Germany, a very wealthy country. It may be seen that the share of young people is much lower than in the world population. Demographers may precisely determine demographic trends many years ahead. The graph presents the forecast age pyramid for the year 2025. The youngest population will decrease rapidly and the population of the elderly, especially women, will rise tremendously. Conclusions from these changes are obvious: we need to develop products and services that will be attractive for elderly people.

Family Life Cycle

The segmentation of potential clients according to their family status is very important for agritourism, because it is families that are the most frequent consumers of these services. There are many possible divisions concerning the family life cycle, e.g. according to Santrock (1998) the stages of the family life cycle are:

- Independence;
- Coupling or marriage;
- Parenting: babies through adolescents;
- Launching adult children;
- Retirement or senior years.

An older system, more commonly known to marketing professionals, is the demographic classification of the family life cycle that classifies them into seven groups as in Table 11.1. The married couples earning two incomes are very important for providers of agritourist services, because they have a considerable income at their disposal, which they use not only to satisfy their basic needs. Table 11.1 and the following description present the American demographic division of families where both spouses have jobs. This has practical importance for the entities providing agritourist services.

What emerges from the research conducted in the United States is the fact that 'full nesters', i.e. couples raising children aged 6–17 years, are the largest segment within couples with a double source of income. As a group their income is average. They are careful and practical consumers: they spend their free time with their families and like material goods and family trips. It is the segment of eager agritourist consumers.

The second largest group is 'crowded nesters' living with their children aged 18–24 years or younger. These couples gain the highest income though they do not have the greatest purchasing power as they support their children in schools and colleges. They may represent different lifestyles within one household. This segment seems to be less interesting as potential consumers of agritourist services.

New parents raising children at kindergarten aged up to 6 years are the third largest segment. The income of these couples is average, so money is a serious constraint for most of them. In these families many women graduated from a college and postponed having children until the moment when they had established their positions at work. In these households comfort and quality become

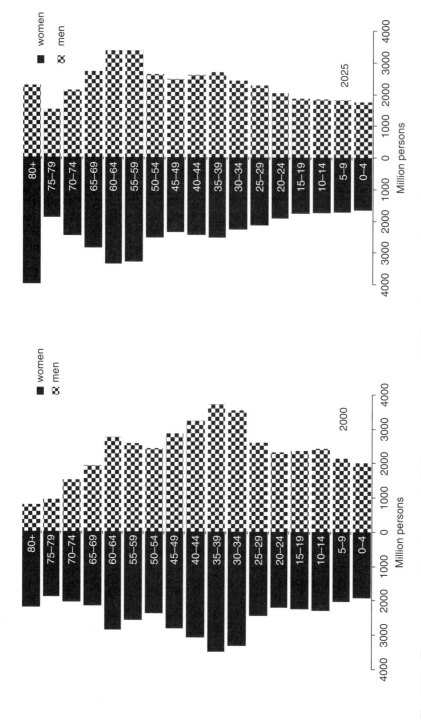

Fig. 11.2.   The age pyramid for Germany in 2000 and forecast for 2025 (from US Census Bureau Current Population Survey, 2007).

Table 11.1. An American demographic classification of married couples with a double source of income, 1991. (From Senauer *et al.*, 1991.)

Name of segment	Percentage of US families with two sources of income
Children at home	
Full nesters	24
Crowded nesters	18
New parents	15
Young families	10
No children at home	
Empty nesters	13
Honeymooners	13
Just a couple	7

increasingly important attributes of goods and services as their income grows. This segment may be the main consumer of agritourist services. Farms with young animals may be especially popular with them.

Young families, the smallest segment, have one or more children aged under 6 years and one or more children aged 6–17 years. These families' income is average. It is usually the wife's income that lets the family be classified as middle-class. Their purchase list includes larger houses, two cars and video and recreational equipment for the family. These families also save money to educate their children. Similarly to the former segment, this one may also be frequent consumers of agritourist services.

Of the couples with two sources of income living without children, the biggest groups are 'empty nesters' and honeymooners. The empty nesters aged 50–64 years have finished raising their children. Some of them have a high income, the others' income is usually average. The wealthier ones invest, buy another house and, when they retire, start travelling. Other couples try to save some money for the retirement years and simultaneously support their children in payment for their studies, purchasing a house or starting new businesses as well as supporting their old parents, paying for medical care, for example. These couples are eager money savers and buy only those products and services they will need when they retire.

Another big segment is young married couples aged 20–30 years whose income is average. The second source of income is important for buying a house and satisfying other needs. The earnings of young married couples do not let them save money, and many of them struggle to ensure economic stability for their families. They try to enjoy the attractions of life before they decide to have children, e.g. they eat in expensive restaurants, go to concerts and travel. They begin to show loyalty to the brand of a product.

Childless married couples are the smallest of the listed groups. The people who make up this segment derive from the final period of the demographic boom. They are aged 35–49 years. They are usually wealthy, set the direction of their lives themselves and live for themselves. Some of them might still decide to have children in the future. Quality is the most important characteristic in all their actions and purchasing. All of the three segments listed above are probably not the best consumers of agritourist services.

Family segmentation data in different countries and at different points in time would identify different groups of relevance. This shows the importance for such segmentation work by the agritourist industry in every country.

The City and the Country

Another important segmentation criterion is the place consumers come from. It is important whether somebody comes from a big city, from a medium-sized or small town or from the country. One may suppose that the people who come from villages and small towns will not usually be agritourist clients. Agritourism is primarily oriented towards city inhabitants, especially from big cities and medium-sized towns.

From the point of view of agritourism it is important to learn the characteristics of city inhabitants. The city population can be divided into the population of small and medium-sized towns and that of big cities.

Table 11.2 presents the development of cities in the world depending on their size. The biggest group is cities of up to 500,000 inhabitants. The total population in these cities is increasing, and their share is over 51% of the urban population. However, their percentage in the urban population is decreasing. The most spectacular population growth is observed for the population of cities of over 5 million inhabitants. The development of the urban population indicates that the potential demand for agritourist services is huge. Enhancing this demand may result in the agritourist farms being unable to cope with this level of demand. This will cause high prices for agritourist services and the necessity to organize mass agritourism.

The bigger the size of a town or city, the higher the percentage of inhabitants interested in agritourism. There are at least two reasons why smaller town inhabitants are not the best clients for agritourism. The first is related to the easy access those people have to agritourist space. The other results from the fact that the inhabitants of towns are often less wealthy than city inhabitants. Thus, the inhabitants of towns populated under 25,000 are not a promising segment for agritourism. Of course, clients from this segment may also appear; however, it will not be the dominant segment. The inhabitants of medium-sized towns populated up to 50 thousand are a better segment. Some inhabitants of these towns may be consumers of agritourist services. However, most clients of agritourist farms will come from big and medium-sized cities.

Education

Segmentation according to education is also relevant to agritourism. It has been shown many times that education completely changes people's lifestyles. The main consumers of agritourist services tend to be people with university or higher education and their children. These people are interested in the world and nature.

Table 11.2. World's urban population in thousands, number of cities and % of urban population by size class of settlement, 1975–2015. (From World Urbanization Prospects: The 2005 Revision, United Nations Department of Economic and Social Affairs/Population Division).

	1975	1980	1985	1990	1995	2000	2005	2010	2015
10 million or more									
Number of agglomerations	3	4	7	10	13	17	20	21	22
Population in urban areas	53,185	69,249	104,507	144,875	183,795	239,655	292,593	326,655	359,238
% of urban population	3.5	4.0	5.3	6.4	7.2	8.4	9.3	9.4	9.4
5 to 10 million									
Number of agglomerations	15	20	21	21	22	28	30	31	39
Population in urban areas	117,232	149,531	152,932	151,605	161,113	193,583	204,492	219,436	272,960
% of urban population	7.7	8.6	7.7	6.7	6.3	6.8	6.5	6.3	7.1
1 to 5 million									
Number of agglomerations	163	197	223	262	295	335	364	422	460
Population in urban areas	316,796	366,800	418,255	494,576	567,733	635,867	713,201	835,440	910,092
% of urban population	20.9	21.1	21.1	21.8	22.3	22.4	22.6	24.0	23.8
500,000 to 1 million									
Number of agglomerations	242	261	299	333	368	403	455	480	494
Population in urban areas	169,971	180,653	206,194	230,362	256,772	278,271	318,204	333,118	346,789
% of urban population	11.2	10.4	10.4	10.1	10.1	9.8	10.1	9.6	9.1
Fewer than 500,000									
Population in urban areas	858,685	970,067	1,101,787	1,249,451	1,381,950	1,497,425	1,621,960	1,759,922	1,929,945
% of urban population	56.6	55.9	55.5	55.0	54.2	52.6	51.5	50.7	50.5

Table 11.3.   Educational background of US population in 2000 and 2007 (in thousands).
(From US Census Bureau, 2008.)

Level of education	2000		2007		2000 = 100
Total	175,230	100.0	194,318	100.0	111
None	851	0.5	839	0.4	99
Primary	27,003	15.4	26,903	13.8	100
High school graduate	58,086	33.1	61,490	31.6	106
Some college, no degree	30,753	17.6	32,473	16.7	106
Associate's degree, occupational	7,221	4.1	8,908	4.6	123
Associate's degree, academic	6,471	3.7	7,862	4.0	121
Bachelor's degree	29,840	17.0	36,658	18.9	123
Master's degree	10,396	5.9	13,607	7.0	131
Professional degree	2,586	1.5	3,090	1.6	119
Doctoral degree	2,023	1.2	2,487	1.3	123

They are ready to spend money to develop their children. Probably very few clients will be recruited from the group of people with primary or vocational school education. People with very low education are usually less mobile and less open to the world. In fact, sometimes they are afraid of it. Nor do they see the reason to learn about the country and agriculture. On the other hand, one may prove that it is the people with lower education that are the right segment for agritourism to address. As these people are poorer, they need cheaper forms of pastime and agritourism continues to be such a form in many countries. Table 11.3 presents the population of the USA according to level of education in 2000 and 2007. One could assume that people with at least graduation from high school, who make up 80% of society, are the segment of potential consumers of agritourist services.

One should note that there are more and more young people learning in schools and studying at universities. This is a positive sign, because in the future agritourism will be able to address a larger segment of educated people.

Income

Another criterion of segmentation that must be taken into consideration is income. Data from the statistical yearbook of the majority of countries show the distribution of people's income in a particular country. Low income automatically excludes a group of potential consumers of agritourist services. Table 11.4 presents the structure of the population of the USA according to disposable income per person in 2000 and 2006.

Although the society in the USA is one of the wealthiest in the world, there is still a large proportion of the population, i.e. as much as 40%, that have annual

Table 11.4. Per capita income in the USA in 2000 and 2006. (Based on US Census Bureau, 2006.)

Salary groups	2000		2006		2000 = 100
Total with income or loss	196,957	100.0	208,491	100.0	106
$1 to $19,999	91,857	46.6	83,413	40.0	91
$20,000 to $39,999	56,448	28.7	58,453	28.0	104
$40,000 to $59,999	25,421	12.9	31,382	15.1	123
$60,000 to $79,999	11,726	6.0	15,764	7.6	134
$80,000 to $99,999	4,552	2.3	7,093	3.4	156
$100,000 and over	6,952	3.5	12,386	5.9	178

income below 20 thousand dollars. Poorer people will never be good customers of agritourist operations. However, the number of people having good and very good income is growing considerably. Currently there is no analysis that shows how the income level of a family affects the demand for agritourist services. Because of a very low income, at least 37% of the US population cannot afford to avail themselves of agritourist services. People must reach a certain income level to be ready to use agritourist services. The income accumulated by consecutive generations also has relevance. If a family is on its way to wealth but they do not own property or a car, they may not be willing to use agritourist services despite their high income. They will probably save their money for the necessary material (house, car) and non-material investments (education). Therefore, the inherited material status is important. A family that has collected material assets may use agritourist services even if they have a lower income. They are more likely to use agritourist services than a family on their way to wealth. Families on their way to wealth are not the best clients of agritourist farms. A large part of the population is still on its way to wealth – hence, they are not very willing to use their finances for recreation. In the future this situation should slowly change in favour of agritourism.

Lifestyles

For providers of agritourist services, classification according to consumers' lifestyles should be a helpful component of the marketing strategy when making decisions concerning the products and services offered and their promotion. When the consumers' needs and wishes are satisfied or fulfilled, the new product or service has a chance to be accepted.

The techniques of classification of lifestyles differ depending on the manner of segment identification. Spatial-demographic techniques, which combine geographical and demographic information, are used to characterize a population according to their places of residence, income, expenses, age and other qualities characterizing tourists in a spatial configuration. These characteristics are used to write descriptions of lifestyles, which provide information about consumers' financial assets and behaviour in a specific geographical region. Psychographic techniques combine the principles of psychology and demography; include socio-psychological, political, theological and economic factors, in the individual's description. Psychography helps describe, explain and classify the changes and interrelations in the values and lifestyles of individuals. The systems of classification of lifestyles are a generalization and therefore they should not be expected to be detailed descriptions of particular tourists. They can be both systems referring to consumers in general and specific systems classifying behaviours related to agritourism.

Research on values and lifestyles show that people's values change as they gradually reach their goals and maturity, that the values and lifestyles change more frequently and in a wider range within some people's lifetime than within other people's lifetime, and that the values affect human behaviour. The values include the complete collection of personal behaviours, beliefs, options, hopes, anxieties, prejudices, needs, tastes and aspirations that together affect human behaviour. Lifestyles are determined by these values.

The simplest segmentation is to divide the population of the country into those who use and those who do not use agritourist services. American studies in the state of California showed that as much as 58% of respondents were interested in agritourism (versus 65% interested in nature tourism) and as much as 68.4% of respondents used agritourist services (Jolly and Reynolds, 2005). The latter can be divided into those who intend to use agritourism in the future and those who do not. Depending on the time of staying on the farm, agritourist clients can be grouped into the following segments:

• Momentary agritourists – spending up to 3–4 hours on a farm, e.g. observing farm work (feeding animals, the milking process, field work, etc.);
• One-day agritourists – spending one whole day on the farm without staying overnight, e.g. having an all-day picnic;
• Agritourists staying overnight – spending all day on the farm and staying overnight;
• Weekend agritourists – arriving on Friday or Saturday and leaving on Sunday;
• Holidaymakers – spending their holiday on the farm (a week, 2 weeks, a month or even longer).

Some tourists visit the same farm several times a year or for several years running. Such tourists are called 'loyal tourists', though their loyalty is limited, of course.

Tourists can also be characterized and divided into different types depending on their motivation for travel and the roles they assume during travels (Cooper *et al.*, 2005). These roles may be identified with and taken on according to social conditions. The classification written by Cohen (Fig. 11.3) is an example of such studies allowing for the social aspects of tourism. He allowed for the elements of personality of potential tourists, defined by means of their opinions, expected impressions and motivation for travelling. Cohen suggested a classification system that accounts for people's tourist motivations: from searching for familiar places of a well-known character to searching for new things, discovering new and unknown places. He divided tourists into two basic groups: mass tourists making use of formalized and institutionalized tourism and tourists practising individual, informal and non-institutionalized tourism. According to Cohen's typology, at

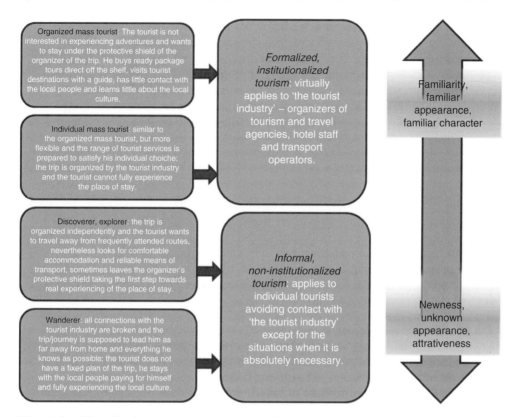

Fig. 11.3. Classification of tourists according to Cohen (1972) (from Cooper *et al.*, 2005).

least two groups of potential tourists, discoverers and wanderers, may be potential clients of agritourist farms and rural tourism accommodation.

Case Studies

1. Based on a statistical yearbook, give the demographic characteristics of your country, analysing:
  a. The size of the population and future trends (development, stagnation, decrease).
  b. Division in terms of age and sex (the age pyramid). Is it a young population or is it getting older?
  c. Place of residence (rural and urban populations, population in terms of the size of urban areas).
  d. Annual income per capita.

2. How do the analysed demographic characteristics affect the development of agritourism?

3. Give a division of the population in your country into agritourist segments.

4. Estimate the volume of the tourist flow and the agritourism-related income flow on the national scale or for a single agritourist farm.

12 Agritourism Markets

The Size of the Market

The market for agritourist products and services is one of many markets within the national economy. International statistics differentiate between the agricultural and tourist markets. However, they do not give data on the agritourist market. Thus it is debatable whether financial results of the activity should be ascribed to the agricultural or agritourist markets.

Demand for tourist services is huge. Table 12.1 lists the size of the tourist market expressed in the travel and tourism demand index for countries with the largest tourist markets, and additionally Poland and New Zealand, i.e. the two countries of origin of the authors of this book. Countries in Table 12.1 are ordered according to the size of the tourist market following data of 2005. The World Travel and Tourism Council (WTTC) uses two economic categories: travel and tourism demand and real travel and tourism demand. The biggest demand for travel and tourism is shown for the USA, with US$1487 billion. Japan ranks second with US$475 billion, followed by Germany, with US$372 billion. The rate of development for the tourism market is also presented in Table 12.1. The fastest rate is found for this market in China, where the index (2005 in relation to 1955) is 385, followed by Russia – 221 – and Australia – 183. A high rate is found for New Zealand, ranking fourth on the list, with an index value of 179. In Japan a downward trend was recorded on the market – an index value of 94.

Apparently no country in the world records detailed statistics for the size of the agritourist market. For the time being, we are not able to specify precisely how big it is. It may only be estimated. The value of this market was assessed in some states of the USA. The state of New York has estimated its agritourism-related income at $210 million and growing (http://www.beyondorganic.com/template/nst.php?id=081705&idy=2005&sn=sn2, accessed April 2008).

Table 12.1. Travel and tourism demand in billions of US dollars in 1995, 2000 and 2005 and the development index. (From [1]World Travel and Tourism Council, Tourism Satellite Accounting Tool; [2]International Comparison Program, 2005.)

	Country	Travel and tourism demand (US$ billion)[1]			Index 1995 = 100			Per capita GDP[a] in 2005[2]	
		1995	2000	2005	2000	2005		PPP[b]	UD$
1	USA	956.58	1358.53	1486.76	142	155		41,674	41,674
2	Japan	503.98	484.07	475.14	96	94		30,290	35,604
3	Germany	291.92	275.52	371.71	94	127		30,496	33,849
4	France	224.34	217.28	314.68	97	140		29,644	34,008
5	Great Britain	183.05	244.16	311.76	133	170		31,580	37,266
6	China	76.89	145.34	296.39	189	385		4,091	1,721
7	Italy	164.32	165.92	228.37	101	139		27,750	30,195
8	Canada	93.08	128.14	165.96	138	178		35,078	35,133
9	Australia	55.31	61.97	101.45	112	183		32,798	34,774
10	South Korea	51.11	52.48	82.89	103	162		21,342	16,441
11	Holland	56.87	60.63	79.57	107	140		34,724	38,789
12	Russia	34.36	29.61	75.84	86	221		11,861	5,341
13	Switzerland	59.08	49.48	69.17	84	117		35,520	49,675
14	Austria	48.18	43.46	67.57	90	140		34,108	37,056
15	Belgium	47.27	47.86	65.39	101	138		32,077	35,852
16	Sweden	28.83	30.69	42.97	106	149		31,995	39,621
17	Poland	20.18	21.50	33.79	107	167		13,573	7,965
18	New Zealand	11.40	10.54	20.42	92	179		24,554	26,538

[a] GDP, gross domestic product.
[b] PPP, purchasing power parities.

The Hawaii Agricultural Statistics Service pegged the value of agritourism-related activities at $33.9 million in 2003, up from $26 million in 2000. There were 187 Hawaiian farms that had agritourism-related income in 2003, and another 145 farms reported intentions of starting agritourism activities. Montana had 10.3 million non-residents visit the state in 2006, spending $2.9 billion.

(Geisler, 2008)

According to a state survey, income from Vermont agritourism totalled $19.5 million in 2002, up from $10.5 million during 2000. Average income received from agritourism for 2200 farms was nearly $8900. In Europe it is estimated that tourists arriving at agritourism farms, rural tourism accommodation and small tourist hotels in rural areas leave about 12 billion euros in them. If we additionally take into consideration the local value added and the multiplier effect of tourism, this value rises to 26 billion euros that stays in rural areas thanks to the development of tourism. However, the estimates more and more frequently quote the total value of rural tourism in European countries as exceeding 65 billion euros. In 1985 the WTO estimates showed that rural tourism comprised 3% of the total tourist traffic. Currently this value is estimated at 10% in Europe (Majewski, 2001). The inhabitants of western Europe devote 28% of their holiday time to rural tourism and it is frequently chosen as a pastime for a second or third time during the year (Sikora, 1999).

Although in many countries it is not yet a significant market, it is regarded as an increasing market. Usually it is easier to achieve success in a rising market. This chapter is dedicated to the agritourist market, comprising the supply, demand and market exchange process.

Demand, Supply, Market Equilibrium

The market comprises all the transactions between sellers and buyers. Originally a market was a specific place where people offering goods met buyers. Today modern means of communication, such as the Internet and telecommunications, and new market tools make it possible to carry out market transactions despite the fact that sellers and purchasers do not meet directly. These tools are especially important in agritourist markets, when the seller and buyer are divided by hundreds or even thousands of kilometres.

Similar to other markets, the agritourist market is the place where service providers, accommodation providers and agents, meet with the purchasers of these products and services. Thus, on the one hand, we have tourists, who want to buy a product or service, and, on the other hand, we have the entities that want to sell them. Demand is generated by tourists who buy products and services, while supply is determined by farmers or other economic entities selling them. Demand is the expression of the will to buy a product or service, and supply is an offer to sell. If there is no supply or if there is no demand, there is no market. There are methods that enable the specification or estimation of demand and supply conditions in the market at the regional and national level.

As in any market, demand and supply are balanced at the market equilibrium point. Since the agritourist market is still at the stage of development, the market equilibrium point is not yet settled. We do not know which consumer segment – in terms of their affluence – is the target segment for agritourism. In the initial stage of development of the agritourist market, very low prices for services defined the market equilibrium point. At that time, it seemed that it was a market for rather poor people. In the course of time, with the development of the market of agritourist products and services and improved quality, the market equilibrium point has been moving in the direction of higher prices. Agritourism is becoming a market for a more affluent segment of consumers. The worldwide prospect for further dynamic growth of urban populations and their increasing wealth indicate that the market equilibrium price for agritourist products and services will continue to rise.

The reaction of the market to an increase or drop in prices and increasing income levels is called elasticity. The more a market reacts to these changes, the greater its elasticity. The elasticity of demand and supply, i.e. the manner in which farmers and agritourists respond to these changes, are an object of analysis. Generally it is claimed that demand in the tourist markets exhibits very high elasticity. Unfortunately, there are no published data concerning the elasticity of agritourist markets.

Instability is the most characteristic feature of the agritourist market, resulting from its seasonality, unreliability and uncertainty. Demand in agritourism is visibly seasonal. Tourists most willingly participate in the holiday season, at weekends and during public holidays. In northern Europe and America, it is difficult to persuade tourists to visit farms during late autumn or winter. However, farmers would like tourists to visit their farms all year round. There have been attempts to overcome the seasonality drawback, e.g. by organizing so-called green schools or opening agritourist restaurants. It may be expected that new products or services will be developed to limit the seasonality of this market. Uncertainty pertains to the transaction itself. Generally tourists are not 100% sure what the quality of the service they are buying is going to be. They will know after they receive it. Similarly, farmers do not know tourists personally and they may not be sure of their behaviour or expectations and frequently they stay in the same house together. Here again the knowledge is better after the transaction.

Another characteristic of the market is its unreliability. Even assuming that the buyer and seller exercise their best will and best intentions, still the service may be unreliable. The causes of unreliability are many: weather, health or simply bad luck. Bird flu causes limitations or even closure of rural areas for outsiders. Watching whales in Kikoura in New Zealand is a classic example of unreliability. If tourists during the sea cruise do not see whale, they will receive 80% of the price back as reimbursement. More examples of unreliability are given in Chapter 8, when analysing risk.

Factors Affecting the Development of the Agritourist Market

The agritourist market is developing constantly. Its development is determined by economic, demographic and social factors and the products and services offered, as well as available infrastructure. This chapter will deal mainly with

economic factors, including gross domestic product (GDP) per capita, the affluence of the society and economic egalitarianism.

The size of the tourist market, including the agritourist market, is determined
first of all by GDP per capita. So far GDP has been expressed in US dollars, but
due to the numerous methodological reservations, we express it in terms of purchasing power parity (PPP). Table 12.1 presents per capita GDP values, expressed
both in US dollars and PPP, in selected countries of the world. The statistics show
that the highest GDP is in the United States and also the citizens of this country
spend the most money on tourist travels. We may observe here a dependence,
namely that countries with a large tourist market are also characterized by a high
national revenue of over 25,000 PPP per capita. China and Russia are excep-
tions in this respect. All the countries analysed here have very well-developed
agritourism, except for Russia and China, where agritourism is only beginning to
develop. There is a very clear relationship between economic indices of PPP for
a particular country and the amount of travel demand for recreational purposes
(Middleton, 1996). Figure 12.1 shows this correlation.

The analysis of PPP alone may not completely explain the rules or motivations people have when spending their money. When researching the income
and expenses of family budgets in Germany, Ernest Engel (Begg *et al.*, 1995)
divided the expenses of family budgets into two groups: food expenses and non-
food expenses. Non-food expenses included all other expenses of households
except food. On the basis of the research conducted, Engel observed that as the
income increases the percentage of food expenses decreases. Thus, the higher
the income, the higher the share of expenses on 'non-food'. Non-food covers
agritourism as well. Figure 12.2 shows the percentage of the income of the

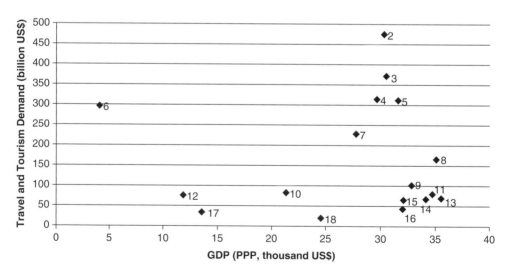

Fig. 12.1. The correlation between PPP and tourism expenditure in 2005. Legend: 1, USA
(outside the graph): 2, Japan; 3, Germany; 4, France; 5, Great Britain; 6, China; 7, Italy;
8, Canada; 9, Australia; 10, South Korea; 11, Netherlands; 12, Russia; 13, Switzerland;
14, Austria; 15, Belgium; 16, Sweden; 17, Poland; 18, New Zealand.

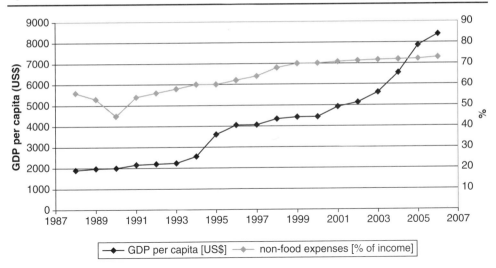

Fig. 12.2. The percentage of income used for non-food expenditure in Poland (based on appropriate Statistical Yearbooks of the Republic of Poland).

population of Poland used for non-food in the last 15 years compared with the rising GDP per person.

Economic Egalitarianism

The market of agritourist services depends on people's wealth. Because the global income of the world population is increasing, so is the expected expenditure on agritourism, other things being equal. However, this does not mean that all people participate equally in the increase. It is important for the sector that as large a number of people as possible should be able to afford agritourist services. Studying the economic divergence of potential purchasers is relevant to the market of tourist services. It is in the interest of tourism that the group of poor people should be as small as possible. In a situation of declining household incomes, tourists choose holidays with their families that are not far away from the place where they live and which are relatively cheap. It is these criteria that agritourist farms and rural accommodation meet.

One of the parameters describing the equality of society's wealth is the Gini coefficient, developed by an Italian statistician, which defines the degree of income dispersion in a population. The Gini value is calculated by drawing a Lorenz curve (Fig. 12.3), which is a cumulative distribution function of the income in the population being examined. If we assume that the total sample population is 100% and the total income of the examined sample is also 100%, we can draw a curve where the X axis shows the percentage of the examined sample and the Y axis the percentage of the total income worked out by this

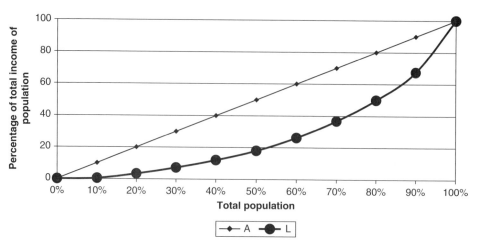

Fig. 12.3. A Lorenz curve.

percentage. Lorenz curves usually assume the form of a fragment of a parabola, where the depth of inflexion of the parabola shows the degree of dispersion of income. The extent to which the curve diverges from the Lorenz line for a hypothetical egalitarian society, i.e. a straight line, can be expressed by the Gini coefficient. It is the quotient of the area over curve L to the total area of the triangle limited by the Lorenz line. The Gini coefficient oscillates between 0 and 1 (or 0–100%), where 0 indicates complete equality of income and 1 a maximum inequality of income. The closer the Gini coefficient is to 1, the less equal is the distribution of society's wealth.

Poor countries have Gini coefficients across the range from low (0.25) to high (0.71), while rich countries generally have intermediate coefficients (under 0.40). The most egalitarian countries are Japan, Slovakia, Sweden and the Czech Republic, where the coefficients are around 0.25. The governments of individual countries may try to prevent the divergence of wealth in society by tax and welfare systems. Denmark and Sweden have managed to lower the Gini coefficient slightly.

The Austrians and Belgians spend the most on tourist travel calculated per person, even though they do not have the highest per capita GDP. This is interesting because these countries have a relatively low Gini coefficient. Where we observe more egalitarianism in wealth, i.e. the lower value of Gini coefficient, the statistical expenses on tourist travel are higher, and this may imply that a greater proportion of society participates in tourist activity.

In many countries in the recent past, the divergence of societies into the rich and the poor has progressed despite an increase in the gross domestic product. Thus, we can observe a situation where an increasing divergence of populations of different countries takes place alongside an overall increase in wealth. This means that societies are growing richer, but the poor are growing richer at a slower rate than the rich. The economic divergence of society means that the majority of the population will use cheap, popular holidays and only a small part will take more expensive holidays. A question arises as to whether agritourism is focused on the bigger but poorer or the smaller but richer segment of society.

Sale of Services and Products

Farmers may use two methods of sale, i.e. direct and indirect. Development of agritourism is followed by the development of methods of sale.

Within direct sales, we talk of both products and services sold by members of the agritourist sector to the end buyers. Products are sold using different methods: in the farm shop, from stands, by 'pick it yourself' or in restaurants. The sale of a service is more complicated. We may distinguish the ad hoc, or spontaneous, sale and sale upon previous booking. The ad hoc sale takes place when an unexpected guest arrives at the farm. Many types of products and services may be sold on this basis, e.g. a visit to a farm shop or restaurant. Sale upon previous booking is effected when tourists earlier inform farmers of their intention to visit the farm. Within direct sales, making them online may prove highly effective. This consists in the agritourist not only booking a service but also paying for it immediately. This system is becoming increasingly popular worldwide. The authors of this study, when visiting agritourist farms in many countries, frequently used Internet booking. Farmers offering agritourist services may encourage tourists to visit their farms also through press advertising. Farmers running agritourist farms supporting so-called green classes may send their offers by post, fax or email directly to neighbouring schools or even visit them directly.

Indirect sale pertains generally to services – especially the booking of accommodation – effected by tourist agencies or agritourist associations. They operate on the market on behalf of farmers. Offices inform farmers of transactions. They charge commission on sales, the amount being substantial at times. These offices undertake actions to form regional and national booking systems for agritourist services on the Internet.

Case Studies

1. Based on statistical data for your country, determine the size of the agritourist market in millions of US dollars, assuming that it constitutes 10% of the tourist market.

2. Determine the income per capita in your country and compare the figures with those from other countries. Is the volume of income in your opinion sufficient for the development of tourism?

3. Check the levels of income for the population of your region or country. Based on this distribution, define the size of the segment that could use agritourism.

4. Propose effective methods of sales for tourist services and products for a selected agritourist farm.

III Agritourist Entities and Enterprises

13 Agritourism Pillars

The third part of this book is dedicated to describing agritourist entities and enterprises, including agritourist farms, food industry enterprises and tourist and transport offices. The enterprises are described in the form of specific examples coming from Argentina, Belgium, Canada, France, Germany, India, Ireland, Italy, The Netherlands, New Zealand, Poland, Russia, Switzerland, the United Kingdom, the United States, Uruguay and Zanzibar. A practical knowledge of agritourism around the world inspired the writing of this part of the book. Visits to various parts of the world enabled not only the observation of various forms of agritourist activity arising but also the evaluation of agritourist chances of regions and individual entities providing these services. At the end of the 20th century, countries seemed relatively monotonous in respect of agritourism. The offerings of farms usually concentrated on horses or holidays by rivers or lakes or in places situated in the mountains. The beginning of the 21st century has brought an explosion of ideas and diversification of agritourist products and services.

The agritourist offer of the world is extremely rich. Farms, food-processing companies and also individuals living in rural areas show extraordinary inventiveness in order to gain extra revenue from tourist service. In rural areas of England, Scotland, Ulster and Ireland the bed and breakfast system is widespread. Providing services for tourists going away for a weekend is of great importance to the rural areas of most developed countries. In France and Italy the PDO (Protected Designation of Origin) concept is used in agritourism. In the United States many farmers try to increase their income through direct sales. They have established associations whose goal is to increase the effectiveness of direct marketing of farm products through agritourism.

Some agritourist farms in the European Union are oriented to receive assistance from the funds of the EU. Others try to gain income from many sources. This section describes a farm generating income from as many as four different sources, i.e. from agritourism, from the direct sale of its products to tourists, from

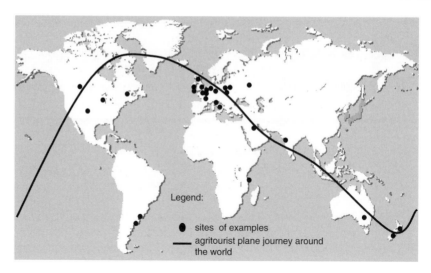

Fig. 13.1. The sites of agritourist examples and enterprises described in this book and the route of the agritourist plane journey.

ecological production and from the European Union funds for unfavoured areas. On the other hand, in such countries as New Zealand, agritourism is developed according to market rules and usually it is oriented towards foreign tourists. Figure 13.1 shows the sites around the world from which examples of agritourist enterprises are given. The continuous line marks the agritourist plane journey around the world described in this section.

Presenting examples of functioning agritourist entities and enterprises is not simple methodologically. They can be described according to various criteria. Therefore this part is divided into five chapters, of which the first three describe the principal elements (pillars) of agritourism: service, accommodation, mobility and agritourist farms.

Chapter 14 is dedicated to facilitating agritourism. The aim is to present the fundamental pillar of agritourism, i.e. information, including maps, guidebooks, flyers and the Internet. The chapter also presents offices dealing with professional agritourism and the problem of the souvenir industry.

Accommodation is another pillar of agritourism. It guarantees satisfaction with an agritourist trip or holiday. This book has already discussed accommodation related to agritourism many times, especially when discussing the products and services of the agri-hotel management. In Chapter 14 we shall pay additional attention to the reservation of beds, their quality and safety. Room reservation provides us with beds in a selected place. A modern system of reservation usually uses emails. We obtain information about the selected place on the Internet and then order it. Reserving a room automatically involves charging our credit card. We can be sure that the reserved bed is waiting for us. It is also possible to look for a bed ad hoc. This option is possible as long as we are not accompanied by little children or impatient people. We can always find a bed somewhere.

When reserving a place to sleep, we believe that the information about it corresponds to reality. However, it is often 'souped up'. This particularly applies to hotels, for which we are usually one-time clients. Particular inconveniences in hotels are street or restaurant noise, loud conversations in the corridor, noisy lift, no drinks and sometimes insects, especially mosquitoes. Many hotels are in league with taxi drivers, who demand from guests much higher prices than usual. Hotels are usually safe places, preventing thefts. However, they aim at providing various extra services in order to increase their earnings.

Chapter 15 is dedicated to mobility, which is a key to the success of agritourism. Means of transport to facilitate mobility are presented with reference to agritourism. The chapter concludes with consideration of the organization of an agritourist trip.

Chapter 16 presents examples of 22 specific agritourist farms, most of which were visited by the authors. Only the description of the greatest ranch in the world comes from the literature. Because of the multitude of possibilities only farms with specific characteristics were selected. These farms usually offered a range of products and services. Their classification was presented in Part II.

Special actions verging on agritourism deserve particular attention. They are on the increase, e.g. agritourism in the world's largest metropolises, farms supporting the educational process in primary and secondary schools – so-called green schools, an incidental trip, forms that become well known due to a particular situation, or even virtual agritourism. All these problems are discussed in Chapter 17, Agritourism on the Edge.

Chapter 18 concludes with a consideration of the predisposition for the development of agritourism in five countries of the world: in Bashkiria (Russian Federation), Canada, the United Kingdom, New Zealand and Poland. Russia is a unique country, with few individual farms. In consequence there is no stimulus propelling the development of agritourism. In spite of this fact, Bashkiria, which is part of the Russian Federation, has a potential for its development. By way of contrast, New Zealand has many owner-operated farms and agriculture and tourism are arguably the most important sectors of the economy. New Zealand, which is remote from larger continents, attracts tourists, offering countless attractions, including numerous agritourist attractions. Canada, an enormous country, has developed a magnificent complex of all-year tourism, at least on the southern confines of Quebec province, along the St Lawrence River. The section on Canada shows southern Quebec as a tourist resort, in which agritourism plays an important role. The United Kingdom, in turn, is a country of many traditions and is agritouristically mature. The section on Poland presents its belt-shaped agritourism space and the detailed space of three regions.

14 Facilitating Agritourism

Tourist Information

The tourist market functions due to information. Today tourists carefully plan their trips either individually or with the guidance of specialists from travel agencies. Clients' need for information depends on many conditions. There is a difference between the information needs of tourists planning individual trips and those of tourists going on, for example, package tours. There is a difference between the need for information before an event begins and during the event. The need for information of tourists on the move, frequently changing their place of stay, are greater than the needs of those going on holiday to one place.

The information provided for tourists must be true, reliable, unambiguous, not misleading and, most importantly, immediate. It is especially important to clearly specify the price and the items included in the service. If the service might involve extra expenses, this must be made clear in advance. The range and manner of preparing information are so diversified that it is hard to classify them.

In order to satisfy the need for information, a very wide range of means and techniques are applied, beginning with oral information, through flyers, brochures and maps and ending with the most modern tools, such as the Internet. In larger tourist resorts and in cities, tourist information offices can usually be found.

Brochures and flyers are usually colourful promotional materials providing relevant tourist information. Their main aim is to advertise a particular enterprise. Flyers are available for free to those interested. The text and graphics of the flyer must be designed in such a way that the most relevant information can be given in a few words and that tourists will be encouraged by the offer. An information flyer must be characteristic so that a tourist in need of information can easily find it and distinguish it from others. Brochures and flyers must be neither too big nor too heavy.

Flyers can be distributed by post, given away to tourists on holiday or to casual people, or they can be displayed on generally available stands in tourist

© CAB International 2009. *Agritourism*
(M. Sznajder, L. Przezbórska and F. Scrimgeour)

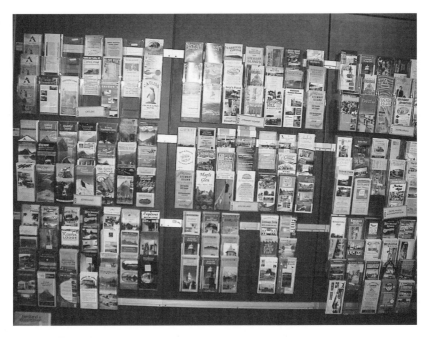

Fig. 14.1.   Examples of tourist flyers for the southern part of South Island,
New Zealand, on display in Invercargill (photo by M. Sznajder).

centres and hotels. The latter form of distributing flyers seems the most effective.
Figure 14.1 presents stands with tourist flyers in a tourist information centre in
Invercargill (New Zealand).

It is very important that flyers and brochures contain only accurate information. Above all, they should include the location and type of farm, kind of production, accommodation, facilities available to guests, distances to, for example, a
bus stop, beach, ski lift or restaurant, and also prices, opening times for guests, the
possibility of arrival with animals, participation in extra attractions and the knowledge of foreign languages. However, the most relevant information is the contact
address and telephone number and directions or a map leading to the place.

Maps and regional guides, which more and more frequently contain offers of
a holiday in the country, make up a very important element of tourist information. Tourists can often get a map with a flyer or brochure free of charge, but they
have to pay for an accurate map of the region. Free maps usually show a scheme
of the location of the place the tourist is going to visit and the way to get there.
In a way they are an introduction to an accurate map. Increasingly, such maps
are published and distributed by communities or agritourist associations. However, the tourist needs an accurate map, which will help him to find and arrive at
the right place.

The Internet is becoming a powerful informative tool in the hands of modern
man. It is also perfect for offering agritourist services. It enables a combination of
all the advantages of the aforementioned means of information. Additonally, it
enables almost direct contact with the tourist and thus a possibility to provide all

the necessary information and answers to questions. The Internet also enables direct sale of services. Although every year the number of Internet users is increasing, it will probably never be the only source of information. Many people, especially the elderly and those less educated with computers, cannot use it correctly and effectively. Nevertheless, in many countries and agritourist farms, e.g. in Italy or New Zealand, the role of this tool has been appreciated and it plays an important role in their promotion.

Pictograms, which are graphic signs communicating more complicated ideas, find increasing importance in practice. Road signs are an important and widely recognized form of communicating information through pictograms. At international airports pictograms communicate information to an international community. Pictograms present information graphically in such a way that every person, regardless of their education or country of origin, can read it correctly.

For example, , the no-smoking pictogram,is clear and unmistakably readable to all people around the world, no matter if a person can read or not. Different kinds of graphic signs that are pictograms of agritourist products are more and more often used in promotional materials. The tourist browsing through thick catalogues of agritourist offers may find it difficult to select the farms on which they would like to spend their free time. Similarly, when arriving at an agritourist farm, they would immediately like to see a board with information about the available services. On the other hand, providers of agritourist products and services want to inform tourists about their services as soon and as well as possible. Pictograms are a solution. Agritourist farms use pictograms to mark the offered products and services. Unfortunately, for the time being there are no standards, though, for example, national tourist federations have developed their own system of graphic signs. However, they are not yet commonly used or identified with agritourist farms. This means that in practice different pictograms are used to mark the same services and, on the other hand, the same pictograms may be differently understood by different people. For example, the pictogram

, used to mark a bike rental, may also be read by tourists as a cycling path. In Fig. 14.2, there are several examples of different pictograms used for the same service by agritourist farms. Hence there is a great opportunity to develop a global, uniform system of pictograms.

Agritourist films can present agriculture and industry in an interesting manner and encourage people to visit the places shown. They should be interesting enough to evoke the wish to see the place having watched the film. It is hard to produce a good agritourist film that tourists would be willing to watch. Agriculture itself shown on television does not attract a very big audience. Both viewers and specialists are reluctant to watch these programmes. On the other hand, a well-made agritourist film can be watched with great interest. For some time there has been observed an increased inflow of tourists to the places where scenes for films were made or to the places described in films. *The Lord of the Rings*, a famous trilogy, is an example. An increasing amount of research is done on the influence of film art on the development of tourism in different regions of

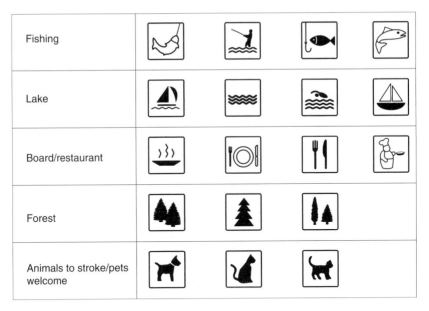

Fishing				
Lake				
Board/restaurant				
Forest				
Animals to stroke/pets welcome				

Fig. 14.2. Examples of different pictograms used to indicate the same agritourist service (from www.agroturystyka.pl).

the world. The analysis made by Frost (2003) based on the films *Braveheart* and *Ned Kelly* is an example.

The goal of an agritourist film is to present agriculture, the country, nature and the production process in such a manner that tourists are encouraged to pay a visit. The film may supplement a visit to an agritourist farm. Tourists are not able to see the whole production process in agriculture, which often takes several months, and in view of this fact the film may bridge this gap. A well-prepared agritourist film is one that presents not only the manufacturing process and attractions but also the beauty of nature varied with a good commentary and music. It is possible to prepare a film where agritourist elements appear only in the background of the main theme. In spite of the fact that they are not in the foreground, they enable communication of the necessary information. An agritourist film may be very short (4–5 minutes, or longer, 15 minutes), or it may even be a feature film. However, the outlay on the latter would be considerable.

An agritourist film should make the tourist interested, show the places they should visit and even point out what is worth buying or consuming. A good film in a way impels the tourist to visit the places shown. Airlines quite frequently cooperate with the tourist business. Often, films are shown on planes before landing in an endeavour to promote the places worth visiting in a destination country. Films may be prepared especially for market segments such as children, elderly people, etc.

The guide serving a group of tourists may as early as on the coach present agritourist films about the places that are the destinations of the excursion. It is best to show the films right before or after visiting a place of interest. Many companies in the food industry invite visitors to watch a film providing basic information about the company before they allow a visit to the premises.

Fig. 14.3. Tourist information office: a railway station converted into the Tourist Information Centre in Blenhaim, New Zealand (photo by M. Sznajder).

Modern agritourism develops with the assistance of tourist information offices. They partly play a role in the agritourist market, where demand, represented by tourists, meets supply, represented by individual employees. In nearly all tourism-orientated towns and in places of tourist interest there are bigger or smaller tourist information offices (Fig. 14.3). Usually the offices are financed by the tourist entities interested in their operation. They may be subsidized by local government or earn income by selling services. A big tourist information office may fulfil many functions. For example, the tourist information office in Wellington, New Zealand, besides providing current information for tourists, has a section with stands with brochures that display flyers from nearly the whole country. There is also the sales department for ferry, coach, train and plane tickets. Sometimes it is connected with an ethnographic or natural history museum. It is also possible to pay for a local trip, book a bed or, finally, buy souvenirs.

It is important for service providers choosing a tourist information office they could cooperate with to know the range of the office's operation, the range of promotional actions and the possible charges they would have to bear in return for advertising.

Quality in Agritourism

The success of agritourist activity depends to a great extent on the quality of services provided because people arriving in the country look for accommodation meeting their requirements. The conceptual framework of quality in tourism was defined and modified by the World Tourism Organization Quality Support Committee at its sixth meeting (Varadero, Cuba, 9–10 May 2003). According to the WTO the quality of services in tourism means:

> the result of a process which implies the satisfaction of all the legitimate product and service needs, requirements and expectations of the consumer, at an acceptable price, in conformity with mutually accepted contractual conditions and the underlying quality determinants such as safety and security, hygiene, accessibility, transparency, authenticity and harmony of the tourism activity concerned with its human and natural environment.
>
> (Improving Competitiveness, 2008)

Veer and Tuunter (2005) emphasize that the current market also makes demands on rural tourism in terms of quality, safety, hygiene and comfort, and in fact the contribution of tourism to rural development is increased by offering products of high quality with a good price/quality ratio.

Uniform standards and rules for serving clients, which inform tourists of the standard and range of services, are a significant factor increasing quality. Hence the categorization and standardization of agritourist facilities and rural tourism become an extremely important instrument of quality evaluation. Thus, the basic goal of categorisation is to ensure an appropriate standard of services by meeting basic requirements below which providing services is unacceptable. The uniformity of service standards is commonly treated as an element of the increasing quality of the offer. On the other hand, it informs the purchaser of the standard and range of the offer, and the seller of the minimum standard of the consumer's expectations. The evaluation systems of the quality of products and services offered by agritourist accommodation applied in most countries emphasise the standard of technical equipment of the accommodation and the elements stressing the rural, agricultural character of the offer. These systems are quite strongly diversified around the world. In the United Kingdom and Ireland the quality of agritourist farms is evaluated by VisitBritain, the National Tourism Agency standard with identical criteria applied for the facilities in the city and in the country. The Quality Department within VisitBritain's Commercial and Marketing Services is responsible for the development and maintenance of the quality schemes, communication of the quality message to consumers and management of the contracts for the delivery of the quality assessment schemes in England. On behalf of VisitBritain, Quality in Tourism assesses over 23,000 accommodation businesses, including agritourism and rural tourism accommodation businesses, across nine different Enjoy England schemes, the domestic branch of VisitBritain (Enjoy England, VisitBritain, 2008).

In France, Switzerland, Austria and Germany the categorization of rural accommodation is carried out by inspectors of national associations of rural/agritourism accommodation providers. The categorization is voluntary and it applies only to the accommodations belonging to the organization. The system of quality evaluation and control is aimed at gaining and retaining the appropriate image of the offered products and services, but it also plays an unquestionable role in marketing.

In France the quality of rural tourism is supervised by a national organization Gîtes de France and the quality sign of agritourist facilities is ears of corn, one to five of which the accommodation may receive (Gîtes de France, accessed April 2008). The higher number of ears marks the higher quality of the services provided. In France camping and caravans are the most popular form of accommodation in rural areas, many of them on farms, but there are also numerous farmers developing camping sites on their farms. Other farmers prefer to invest in various kinds of short-term rental houses, known as gîtes. The quality of the rural tourism product has been improved over time; however, there is no complete product as yet (Veer and Tuunter, 2005).

In Austria a system of quality classification (rating system) was introduced by the Austrian Farm Holidays Association (Urlaub am Bauernhof) in 1993 and modified in 2000 (Urlaub am Bauernhof, accessed April 2008). The facilities (house, interior, lounge, etc.), services (catering, guest information, quality of holiday experience, educational activities, etc.) and authenticity (farm quality, location and accessibility, construction design, garden, exterior, etc.) of agritourism enterprises are checked at least every 4 years by a three-person commission. The emphasis on quality is marked with daisies (flowers), and the best farms receive the maximum of four daisies (Fig. 14.4). The Austrian Farm Holidays Association explains that there are holidays spent amongst the flowers in two senses of the word. Daisies guide clients' way to holiday accommodations and, instead of a three-star hotel, for example, an agritourist will find a three-flower farm. There may also be a house symbol, which is displayed in addition to the daisies, indicating that the establishment is a 'farm guesthouse' with more than

  Two Daisies: Functional amenities in personal accommodation and throughout the farm. WC, shower or bath are generally on floor.

 Three Daisies: Comfortable amenities in accommodation and throughout the farm. Shower or bath and WC generally in the room or apartment. Farms where one can feel totally at ease.

 **Four Daisies**: An outstanding setting from every perspective, with comfortable amenities and quality furnishings. Shower or bath and WC in the room or apartment.

Fig. 14.4. Quality marks in Austrian agritourism enterprises (from Austrian Farm Holidays Association (http://www.farmholidays.com/bundesverband/qualitaet.html?L=&id=1&L=3, accessed April 2008).

30 guest beds. Introduction of the quality system has resulted in a great deal of positive publicity and improved the image of the rural tourism product in Austria (Embacher, 2003). All farms are regularly inspected by a commission that looks at 150 quality criteria. The quality of the farm, amenities and service are all evaluated as part of the process. With the new system of 2000, entrepreneurs can specialize in categories such as ecology, health, children, disabled persons, horse riding, cycling and wine tourism. For a specific category, a business must score between 50 and 100% on the relevant criteria. Of the total number of businesses belonging to the Association, 18% have a speciality. A substantial improvement in the quality of agritourism businesses was observed (Veer and Tuunter, 2005).

In Switzerland the Association of Swiss Holiday Farms (Ferien auf dem Bauernhof) set up a quality label (Ferien auf dem Bauernhof, accessed April 2008). In order to receive a quality sign and the right to use it for marketing purposes, agritourist farms must have at least five species of domestic animals, of which at least one must be a farm animal and they must offer guests at least three products coming from their own farm. This is not just about well-cared-for accommodation, but also about factors relating to the surrounding area. Switzerland offers a good price/quality ratio product in rural tourism (Veer and Tuunter, 2005).

In Poland the legal basis defining categorization requirements for tourist accommodation facilities is the Tourist Services Act of 1997. However, it lacks clearly defined standardisation criteria in reference to agritourist services; it does not even define the term agritourist accommodation and it includes it in other hotel facilities. This is a serious problem in the interpretation of standardization needs and requirements. The minimum standards the law imposes on such facilities include only the meeting of requirements referring to the equipment for and compliance with sanitary, building, fire and other requirements specified in separate regulations. Hence, on the basis of legal regulations a System of Qualification and Categorization of the Rural Bed Base was developed and managed by the Polish Federation of Rural Tourism 'Hospitable Farms'. The quality evaluation of agritourist farms and rural accommodations and their control is supervised by qualified inspectors of the Federation.

The Federation was established in 1996 as an association of regional associations and almost from the very beginning it started to develop a national system of categorization and standardization of accommodations. Owing to the funds from the Phare-TOURIN II project realized in the years 1995–1997, a voluntary system of categorization of the 'rural bed base' in Poland was developed. The companies DG Agroprogress International GmbH from Bonn and Tourism Development International Ltd helped with the implementation of the categorization system.

Standardization requirements were determined for such bed facilities as guest rooms, independent residential units (holiday flats and detached houses), group rooms and accommodations and campsites near farms (for bungalows and tents). For each of the facilities minimum requirements were set out on the basis of which it is possible to evaluate their quality and to mark them with a specific category: from the lowest, Standard, through category I and II to the highest category III. Marking accommodation with a certificate is supposed to increase its competitiveness and the tourist receives information about the place they intend to visit. The owner of certified accommodation has a possibility to present their offer at domestic and foreign tourist fairs at the stands co-financed by the Federation and to place their offer in promotional materials of the Federation.

The evaluation is made by inspectors and certification given for the period of 2 years. Information about the listed farms can be found in the countrywide catalogue or website of the Polish Federation of Tourism 'Hospitable Farms' (accessed April 2008). Undergoing certification involves rather high costs for the owners of agritourist accommodations, because stopping the inflow of assistance means from the European Union has resulted in a considerable price increase and in consequence the categorization process of the rural bed base in Poland has slowed down. What emerges from the data of the Federation is the fact that in the years 1997–2003 as few as 1020 accommodation providers submitted their facilities to voluntary evaluation of the rural bed base. The categorization fee for members is now 250 zlotys for up to five rooms. For non-members it is 600 zlotys. If there are more than five rooms the additional fee of 20 zlotys for each extra room must be paid. In both cases the inspectors' travel costs are calculated as extra costs. Besides the statutory requirements, accommodations must meet general requirements specified by the system. For example, they must be properly marked, keep the driveway into the area, the yard and the buildings in good condition and ensure appropriate lighting of the facility and a suitable temperature in the rooms used. Furthermore, detailed rules were drawn up that specify the conditions that must be fulfilled for accommodation to receive a particular category. The Polish certification system was drawn up on the basis of existing and proved certification systems from European countries, but it is also similar to the systems used in non-European countries, e.g. in New Zealand.

In New Zealand the categorization of agritourist accommodation and rural tourism is done by the New Zealand Association of Farm and Home Hosts '@ home New Zealand' parallel with the Qualmark® company, which evaluates the whole bed base (@ home New Zealand, accessed April 2008). However, it turns out that the quality requirements set by Qualmark® are frequently too high to be met by small accommodation providers. In view of this fact for quite a long period of time negotiations have been held between the Association and the Qualmark® company to adjust Qualmark®'s quality requirements to the possibilities of small tourist accommodation.

Based on the best practices and experience of various countries, EuroGites the European Federation for Farm and Village Tourism, has developed standard criteria for quality assessment standards for rural accommodation (EuroGites, accessed April 2008). They include the following main blocks: equipment, surroundings, services in the accommodation and in the surroundings (<15 km), personal attention, privacy, ambience and security. Detailed standards are available at the EuroGites website. In each block there are several rules describing minimum requirements both for the domestic market and for the international market. The standards are not compulsory, but there is a 'strong recommenda-

tion' for members of EuroGites. The standards complement the existing classification, regulations and quality systems in European countries. According to the
recommendation of the Federation, they should be respected in the case of marketing specific accommodation units under the name of the organization. However, each member organization is responsible for defining the extent and manner
of application of the standards in the area of its responsibilities. Inspection and
control of the rules are under the supervision of respective member organizations, but the European Federation reserves the right for its own controls (Quality
Assessment Standards, 2005).

The European Commission, focusing on rural tourism and aiming at those
concerned with the management of rural areas as tourism destinations, supports
the use of integrated quality management (IQM) to ensure that visitors have a
special experience, meeting or exceeding their expectations, while maximizing
the benefit to the destination. In 2000, the European Commission published a
paper on Integrated Quality Management in Rural Tourism Destinations: Towards
Quality Rural Tourism: Integrated Quality Management (IQM) of Rural Destinations (2000).The EC paper sets out a code of practice for integrated quality
management in rural tourist destinations that is based on the experience and
success factors emerging from the case studies specifically written for organizations responsible for tourism in the destinations. The IQM paper introduces
'fifteen principles' in the Code of Practice, namely integration, authenticity, dis-
tinctiveness, market realism, sustainability, consumer orientation, inclusiveness,
attention to detail, rationalization, partnership, interdependence, time, commit-
ment, accurate communication and monitoring. The IQM paper stresses that the
principles are relevant to private enterprises, as are many of its recommendations
of good practice. Although the IQM is not compulsory for rural tourism and
agritourism enterprises, the code of practice for IQM is recommended for rural
private-sector enterprises (Integrated Quality Management in Rural Tourism Destinations, 2000).

Professional Agritourism

All over the world there are special travel agencies established specializing in sup-
plying services to people interested in agritourism and staying in the country.
There are many such agencies and bureaus in countries where staying on farms
and in rural areas has long been a tradition. New Zealand is an example of a
country where there are several agencies serving tourists interested in staying in
the country. A comparison of such agencies and their offers is shown in Table 14.1.
AgriTours Canada Inc. is an example of such an agency. The logo of AgriTours
Canada Inc., together with the trademarks of New Zealand and Polish agencies,
are shown in Fig. 14.5. AgriTours Canada Inc. is a professional travel agency
serving mainly farmers and agricultural experts, whose aim is to visit farmers in
other parts of the country or the world. This agency provides services in proper
agritourism. During the year the agency organizes study trips to Canada for
about 80 groups from all over the world, chiefly from Europe. The goal of these
study trips is to present the production of plants, milk and herds in Canada.

Table 14.1. A comparison of some tourist offices and travel agencies organizing stays and tours on agritourist farms in New Zealand. (From Hospitality Plus; Rural Tourism Holding; Agritour; Farm to Farm New Zealand Tours; Agri-Travel International; New Zealand Farm Holidays Ltd; Agrotravel.)

No.	Name of company	Seat	Offered tourist products and services
1	Rural Tourism Holding – RTH	Cambridge, North Island	Organizes farmstays, nights and stays on bed-and-breakfast farms and various attractions on farms; trips for schoolchildren, students or individual tourists, also specialized visits to farms.
2	Rural Holidays NZ Ltd	Christchurch, South Island	Sells rural holidays with farmstays or countrystays, organizes activities on farms and in rural areas, organizes group stays in the country.
3	Farm to Farm New Zealand Tours	Rangiora, Christchurch, South Island	Offers visits and specialized tours of farms and orchards for interested individuals and groups (including institutions), accommodation and holiday farmstays and countrystays, organizes transport and guided tours in rural areas.
4	Agri-Travel International (a branch of Calder and Lawson Travel)	Hamilton, North Island	Offers package tours in the country, adapts them to tourists' needs and requirements, especially specialized visits concerning agricultural production, including accommodation, board, transport, guided tours, also in foreign languages, and sightseeing.
5	Hospitality Plus, the New Zealand Home and Farmstay Co. Ltd	Gerladine, South Island	Offers short stays and holidays on farms, in New Zealand homes, and also language learning on New Zealand farms.
6	Agritour, New Zealand's Specialist Agricultural Tour Company	Hamilton, North Island	Sells packages comprising visits to specialist research institutes dealing with problems of agriculture, agri-business and forestry, to various farms, transport, board, accommodation, sightseeing, guided tours and courier service.
7	New Zealand Farm Holidays Ltd	Kumeu, Auckland, North Island	Offers farmstays, New Zealand homestays, nights and B&B in various regions of New Zealand for individuals, small and large groups, sells open tickets – vouchers for stays and beds on various farms, organizes dinners in the homes of New Zealand families.
8	Agrotravel	Suchy Las, Poland	Prepares tailor-made agritourist trips to Poland for foreigners according to visitors' expectations.

Fig. 14.5. The trademarks of travel agencies serving professionals interested in agritourism and rural tourism (from Hospitality Plus, Agri Tours Canada Inc., Agrotravel).

Apart from the educational aspect, the trip programme also includes other items related to the culture and tradition of Canada. Farmers have a chance to see and visit interesting places, but they also have time to relax. Running a travel agency concentrated on proper agritourism involves high financial risk. For example, the foot-and-mouth disease epidemic in western Europe had a very negative effect on the financial results of bureaus and agencies.

In view of the fact that in some countries there are no professional travel agencies specializing in proper agritourism, ad hoc trips are prepared by universities, the staff of consultancy centres, consulting companies, etc. Professionals in many branches when visiting a particular country very often use the service of the staff employed by these entities, because they are the only ones that can provide cheap and professional information about the agriculture and industry of a particular country. Agricultural universities themselves are also a tourist attraction as they are frequently situated in former estates, where sometimes agricultural production still continues. For example, this applies to agricultural universities in Rennes, Thiverval-Grignon and Montpellier in France, the Scottish Agricultural College (SAC) in Scotland, Plymouth University in the United Kngdom, the Warsaw Agricultural University, etc. Many of these have rich traditions and they frequently take pride in the fact that distinguished people lived and worked there. For example, the Agricultural University of Montpellier is proud of the fact that Louis Pasteur, the great French chemist and microbiologist, presented his lessons in the lecture hall which carries his name today.

The role of tourist offices may be evaluated in different ways in the future. On the one hand, it is individual rather than mass tourists that take more interest in agritourism and rural tourism. From the very beginning they look for direct contact with the farms offering tourist services in the country without any agents whatsoever. However, on the other hand, professional offices sell whole packages of products and services, comprising not only beds but also board, transport, sightseeing, insurance, various extra attractions, etc. Therefore using the services of such agencies may be very convenient and may save potential tourists or excursionists a lot of time or even money.

The Souvenir Industry

The goal of the souvenir industry is to manufacture souvenirs and sell them to tourists. More and more frequently the souvenir industry concerns agritourism also. Sometimes souvenirs are given away to tourists for free. Then they have an advertising character. The souvenir industry gives important support to tourist activity. Agritourist elements, breeds of animals, old buildings are perfect models for the manufacture of souvenirs. Their diversity is enormous now, starting with T-shirts, through postcards, maps, albums, car number plates, wooden toys and folk clothes, and finishing with small gadgets. Also, these may frequently be national products: tinctures or sweets, etc. For example, in souvenir shops on the island of Texel one can purchase souvenir Texel sheep made of wood, coated with sheepskin, etc. Wooden lambs are of different sizes and they are sold at different prices, from 2 to 100 euros. The Texel breed brings multiple profit for the local people: first of all, as a farm animal, secondly, as a tourist attraction and, finally, as a model for the souvenir industry. In nearly all countries we may notice that specific species and breeds of animals are frequently used as models for the souvenir industry. In Canada, elk, wolves and bears are used as models, in New Zealand kiwis and moa birds, in India elephants and peacocks, in Poland on the Hel Peninsula Baltic seals, on the island of Jersey, Jersey bred cows, etc. The distribution of souvenirs is usually the responsibility of special shops located in tourist centres. Usually they are situated in attractive city streets and squares, at tourist information centres and also at airports in the departure section. In small amounts souvenirs are also offered on farms and at agritourist entities. Usually these are farm-made products. Figures 14.6 and 14.7 show examples of souvenirs with agricultural and natural main themes.

Fig. 14.6.   The souvenir industry: Hungarian national fabrics (photo by M. Sznajder).

Fig. 14.7. The souvenir industry: street stalls with national and religious souvenirs, Vadodara, India (photo by M. Sznajder).

Case Studies

1. Collect agritourist brochures and flyers from your region and country and think if they describe the agritourist attractions of the area.

2. Explore the Internet for the availability of information on the agritourist attractions of your region or country.

3. Think if the agritourist farms you know use pictograms and if they are clear.

4. Probably your region is short of agritourist souvenirs that tourists could buy. Think of the kind of souvenirs that could be manufactured, at what prices they could be sold and how big the demand for them could be.

15 Mobility in Agritourism

Means of Transport

Mobility, that is, the ability to move, is an important element of agritourism. Without the possibility to move, tourism cannot exist. This also applies to agritourism in a specific way. Modern civilization has provided humanity with fast and comfortable means of transport. In antiquity and the Middle Ages people's mobility over long distances was possible owing to horse-drawn carts and ships. Because of their slow speed and small capacity only a few people were given the possibility to travel. Mobility was significantly facilitated by the development of the railway, then by the car industry and in the last century by aviation. The development of fast means of transport has resulted in a considerable shortening of the journey and this has facilitated the development of tourism all over the world. Within 1 hour, a plane travels a distance ranging from 500 to 1100 km, which corresponds to 10–15 hours of a continuous car journey and to about 150 hours of a horseback journey. In former times a sea voyage from Europe to New Zealand lasted about 3 months, whereas today we can travel this distance by plane within 1.5 to 2 days (27–33-hour flight). Currently the most frequently used means of transport in both tourism and agritourism are the car and coach. However, these are not the only means of transport. Other means of transport used in tourism are horseback (or by camel, donkey, elephant), bicycle, motorcycle, railway, boat, ferry, helicopter and plane. Not only does using these means of transport involve the speed of travelling but also each means has its unique atmosphere.

The tourist may move on foot but this limits mobility. It is characteristic that the tourist blends in with nature. They stay longer in the country; have a better opportunity to learn about elements of agriculture and more contact with nature. Each day they can walk a distance of 20 to 45 km. Walking tours are usually taken along marked routes, in the areas of interesting natural or scenic values

Fig. 15.1. A walking tour in the Bieszczady Mountains (photo by M. Sznajder).

(Fig. 15.1). These are usually one-day excursions and sometimes walking camps, which last from 1 week to several weeks.

In Poland, going on foot, even for long distances, is also related to pilgrimage traffic. Walking tours are frequently only part of the whole journey. Combined trips are organized, i.e. part of them is on foot and part by other means of transport. More and more frequently a car accompanies the pilgrims. Pedestrians walk along a marked route and the car transports their luggage to the place of another stopover or where they will stay overnight. In agritourism moving on foot is related to educational agritourist paths crossing the rural landscape. While moving on foot is not supposed to be arduous, it involves the need to wear proper clothes and, especially, to wear appropriate shoes.

Moving on horseback may be either a tourist attraction or a necessary means of transport. If an agritourist farm organizes trips on horseback, it is a tourist attraction. Tourists visiting the great pyramids in Giza are offered a ride on a camel. Likewise, in many places children are offered a ride on a pony. Visiting many parts of the world without riding on horseback is not possible. If somebody wants to visit the immense pathless tracts of Russia or Mongolia, they have to travel on horseback. Similarly, in Africa cultural tourists often ride camels to reach corners that cannot be reached otherwise. Riding on horseback or camelback is thus necessary in the case of tourist adventures to distant, pathless corners of the world.

A bicycle is convenient and in some situations the only means of ensuring mobility in rural tourism and agritourism. In many areas there are cycling paths built to make trips that last even several days. Cycling paths may link agritourist farms or there can be a network of paths within a farm. A farm of over 500 ha may make a network of cycling paths by itself. In such cases the tourist usually travels on their own bicycle. In some places bicycle rentals are indispensable and they are very popular. In particular this applies to small, very attractive areas such as the islands of Jersey (Box 15.1) and Guernsey in the English Channel or Bornholm in the Baltic Sea. When organizing a group trip where part of the journey is made by car and part by bicycle it is possible to use mobile bicycle rentals.

The goal of a mobile bicycle rental is to provide organized groups of tourists with bicycles in the places where tourists expect them. A mobile bicycle rental consists of a van or bus and a trailer to transport bicycles. Bicycles may be rented for a short ride or for a trip that lasts several days. Tourists may choose to ride a bicycle along selected parts of the journey and they may travel by other means of transport when the route of the journey is less attractive.

A motorcycle is not a very popular means of transport in agritourism. There are reports of motorcycle trips across continents, but these are more frequently sports feats rather than tourist trips. Besides, there are far fewer tourists holding a motorcycle licence than a driving licence.

One type of motorcycle is a rickshaw – a cheap means of transport in Asia (Fig. 15.2). It is this means of transport that may enable excursions to rural areas.

The motorcycle is often a popular means of transport on farms with a developed system of paths and lanes in them. It very often supersedes horse transport and is used even to transport animals (Fig. 15.3).

Box 15.1   A bicycle tour

The island of Jersey (area 117 km^2), which is slightly oblong in shape, is 12 km long and 9 km wide. Politically it is a British dependency. Above all, this is a seaside resort and also a perfect place for rural tourism and agritourism. The island is well known all over the world, because it is from there that a popular breed of dairy cattle comes – the Jersey. It is also a place where flowers are produced. The means of transport that makes it possible to move around the island is the bicycle. Because the island is not big it is possible to reach every corner by bicycle quite quickly. On the island there are numerous cycling paths, which are called 'green lanes'. Most tourists arrive on the island by fast ferry, yacht or plane. The tourist without a car looks for a convenient means of transport, which the bicycle proves to be. The deposit to rent a bicycle is about £150. For £8 one can use the bicycle all day. Many tourists leaving the central ferry terminal in St. Helier go straight to one of several bicycle rentals situated next to the ferry terminal. On Jersey, as in Britain, one must drive on the left. The inhabitants show much understanding for 'the Europeans' riding bicycles on the right: they do not always remember to ride on the left side of the road. A bicycle rental enabled one of the authors to make a trip around rural areas of the island. The goal of these trips is usually to observe and photograph Jersey cows in their natural habitat.

Fig. 15.2. Rickshaw three-wheel motorcycle – popular transport medium in Mumbai, India (photo by M. Sznajder).

Fig. 15.3. Motorcycle used for dog transport on a hilly farm in New Zealand (photo by L. Przezbórska).

The passenger car has become the main means of mobility all over the world. However, there are immense areas of Africa, Asia and South America where it is still a rarity. Usually people using agritourist services arrive at the place in their own cars. Weekend tourism or even occasional weekday tourism usually requires the use of a car. The type of car depends on the consumers' wealth. Some tourists go caravanning to be independent of looking for or booking beds. There are many advantages and disadvantages of this means of transport. Undoubtedly poor agility and slow travelling speed are the disadvantages.

At airports there are car rentals operated by such companies as Avis, Hertz, Alamo, etc. A car rental makes it possible to combine a plane and car journey.

Some trips require the use of off-road vehicles that travel easily along pathless tracts and across undeveloped areas. Drivers sometimes use the beds of shallow rivers as roads, or they travel along pathless tracts in the mountains. A journey in an off-road vehicle on mountain slopes may give an exhilarating experience to the tourist, especially if the area has steep slopes. Many times the authors have had opportunities to travel in off-road vehicles in the mountainous areas of Scotland, the Urals or New Zealand. Such journeys always make for exhilarating and unforgettable experiences.

A coach is a very convenient means of transport for a group of people. Usually a coach with a driver is hired from a transport company, though there are also van rentals where it is the client's duty to supply the driver. Coach standards are very diverse. On longer journeys by coach, it is necessary for the coach to have air conditioning, a toilet and a microphone, as well as a set of monitors to watch films. It is important that the windows of the coach should enable easy observation of the terrain. Quite frequently there are coaches whose suitability for agritourism is doubtful, especially if the windows are so small that they make watching the views difficult.

The train is still an important means of transport for tourism. The development of railways above all depends on there being sufficient travellers willing to use its services. If there is a high population density in a particular country, this system develops very successfully. If the population density is low and few passengers use this means of transport, the railway declines as an expensive means of transport. A low population density in some regions of countries caused the importance of the railway as a means of transport to decline quite rapidly. Narrow-gauge railways have declined almost completely. In some places museums and short sections of narrow-gauge railway have remained as a tourist attraction. At present, the train is often considered to be a means of transport for wealthier people. In western Europe rail is still a very popular means of transport. In China and India rail will probably never lose its importance.

Travelling by rail requires attention. The critical moments of a train journey are arriving at the station, buying a ticket, taking a seat on the right train in the right carriage and placing the luggage. Tourists often have doubts concerning their safety, especially when travelling at night. It is important to get off at the right station. Having got off the train one has a problem again to reach the destination. In spite of this the train journey is attractive and provides an extra possibility to observe the rural and agricultural areas one is travelling across. To use this opportunity, access to the window is particularly important; for example, it is possible to take many interesting photographs. If one intends to spend the time of travel observing an area to be traversed, a question arises as to whether to reserve seats. Railways offer a number of different ways to reserve and buy tickets, including reservation on the Internet. Experience shows that sometimes a reservation frustrates plans. One can get a seat that does not fully satisfy as it makes it difficult to observe the terrain. Furthermore, if one intends to take a few photographs, requirements concerning the seat not only increase but also change during the journey itself. Sometimes it is better to travel in carriages without reserved seats, where it is possible to choose the vantage point freely. In many cases travel without prior reservation of a seat is not possible. On long journeys that last several days, it is important to buy a place to sleep. The atmosphere of train travel is different in different parts of the world. To illustrate this diversity, Box 15.2 describes a day train journey in India, a 3-day train journey to Bashkiria and a journey on a rapid train in France and an indication of how a traveller can make a train agritourist trip from them.

Box 15.2.   Three train journeys

Anand–Mumbai

It is a good idea to travel by train in India. It is easy to meet people and the view from train windows gives a possibility to form an opinion about agricultural production. A train journey in India is an unforgettable experience for a European. Express trains are long. Thousands of people travel on them. Their difference from European trains is obvious. Above all, they do not seem to be very clean and they are crowded with peddlers selling food and drinks. The journey can be facilitated by buying a ticket electronically. The journey from Anand to Bombay took place in early March, at the beginning of the dry season, when harvest time in the fields is about to end.

The journey of 427 km takes about 10 hours. On the way the train passes villages and huge cities (Vadodara, Surat, Gapi). Almost all the way it crosses flat areas, occasionally cut by river valleys. From time to time dry gorges and ravines can be seen. The closer to the Arabian Sea it is, the more visible is the influence of tides on them. On the way sun-scorched grass shaded with bushes or trees can be seen. There are buffaloes on the dry grass grazing individually or in groups, or hiding in the shade of huge trees or thick bushes. There are banana, tobacco and rice plantations and orchards. There are plantations of sugarcane at different stages of ripening and fields from which cotton has been harvested. Occasionally there are herds of goats driven by goatherds. From time to time there are camels resting in the shade of trees. India provides food for its population. Sometimes it is said that there is even a surplus of it. Unfortunately, slums can often be seen with no water in them, which are a shame to contemporary mankind. Women do laundry and Indians have a wash in flows of water of doubtful quality. However, beautiful houses and temples can also be seen. There is a mountainous area stretching right before Mumbai. Also, close to the city there are piles of salt obtained from seawater. The journey ends at Bolivardi station. There begins an endless stream of people walking in different directions. The train journey, which gave an idea of agriculture, breeding animals and people's living standard, is over.

Journey to Bashkiria

The journey from Poznań to Ufa, the capital city of Bashkiria, takes almost three nights and days. We know the agricultural landscape of Poland – flat terrains, fields of corn or other cultivated plants, scattered villages and households. On the Polish–Belarussian border wheels are changed because the railway track is different here. It takes a long time. We leave Terespol after dark. We shall see the morning in Russia. It will uncover rather poor, wooden cottages and villages. It is evident that they need redecoration. In Moscow we arrive at Belaruskaya Station. The travel from Belaruskaya Station to Kazanskaya Station requires that we use the metro. This may be a problem, especially for those who do not know the Cyrillic script. In the afternoon from Kazanskaya Station we leave for Bashkiria on the Bashkortostan express train. The train is clean, quiet national songs and tunes can be heard, curtains in the windows carry national elements, teacups and biscuits. *Kipiatok* is available on the train all the time. This creates an encouraging atmosphere. The journey from Moscow to Bashkiria is long, but also interesting. From the windows we can see villages and small towns, this time in better condition and wealthier, petrochemical plants, immense agricultural expanses and, above all, great forests and rivers. On the way we cross the wide River Volga. The other bank is far away. We pass Samara. A half-an-hour break in the journey. It is worth catching some fresh air. In Russia there are many time zones. On the train there is Moscow time, outside local time. To avoid misunderstandings resulting from this, public address system announcements at train stations on the way do not tell the train's departure time but inform passengers how much time there is still left until the departure of their train. During the journey travellers strike up numerous acquaintances. People are very talkative. At stations sellers work hard offering the travellers food, usually dried fish. In the restaurant car national Russian dishes are served. We cross the River Belaya and soon it will be time to get off. The following 24 hours of the journey passed very quickly.

Rapid trains in France

In France the TGV rapid railway system and the very rapid Thalys railway operate. A journey on TGV is fast and comfortable. The journey across France from north to south takes about 5 hours. Perhaps rural landscapes change too rapidly while travelling across the country. This makes agritourist observation slightly difficult. In fact, the TGV system is not very comfortable for travellers with bulky luggage. The dynamic development of the system of the very rapid Thalys railway, owned by Belgian, French, German and Dutch railways, indicates that railway transport has a great future. The railway serves Amsterdam, Brussels, Cologne, Ostend and Paris, and soon also Marseille, Bordeaux and Geneva. It enables extremely quick passenger transportation between major European cities. A journey on these trains does not create such an atmosphere as, for example, exotic train travel in Argentina. The train there is a place of continuous sale of anything that can be sold and the venue of artistic performances.

Tourists sometimes also have the opportunity to use water transport. On seas and on large lakes there are ferries plying, and on rivers motorboats (see Fig. 15.4). Travellers may also use rowing boats, canoes and speedboats. Ferry transport is popular on the Baltic, the North Sea, the Irish Sea, the

Fig. 15.4. The agritourist farm situated on Lake Whakatipu, New Zealand, can be reached only by boat (photo by M. Sznajder).

Mediterranean and Indonesian seas, but also in the Caribbean. Modern Linx catamarans sail at a high speed. For example, the journey on a fast catamaran from Buenos Aires to Montevideo, the capital of Uruguay, which is about 210 km away, takes only 2 hours. Sometimes, if the tourist wants to reach the destination, they can only use a boat. For example, sightseeing in the Paraná Delta is possible only by means of a boat. The delta is a tangle of rivers and their arms surrounding a multitude of islands. Contrary to one's expectations, the delta is not a wild uninhabited empty space covered only by the jungle but is the place of residence of wealthier people. Everywhere it is full of life and tourists are welcomed. The system of motorboat transport in the Paraná Delta is well developed and slightly resembles a tram or bus network in a big city. Boats of individual companies travel along marked routes according to a timetable. Usually every household has a landing stage at which boats arrive.

Canoes (and rafts – see Fig. 15.5) enable travel along interesting watercourses. While canoeing we can observe the rural and agricultural areas we pass on the way. In Europe there are excellent canoeing conditions, especially for downstream canoeing. For example, this applies to the River Biebrza and the Bug Canyon, but also to other rivers in Poland. Canoes are rented in the upper course of the river and returned at the appointed place on the lower course of the river.

For wealthy clients who want to visit many interesting places quickly and comfortably, helicopter tourism is offered. It is especially New Zealand that specializes in helicopter tourism, though in Poland it is also possible to have a bird's-eye view of the great Masurian lakes, for example. In New Zealand, where the road network is poorly developed, there are considerable altitude differences

Fig. 15.5.   Rafting for pleasure on an agritourist farm in Poland (photo by M. Sznajder).

within short distances and reaching interesting places in the Southern Alps requires travelling for long hours, helicopter tours are offered. Within 3–5 hours it is easy to see the beautiful fiord Milford Sound, walk on the Fox glacier, fly over lakes and mountain peaks (Fig. 15.6), and see the smoking volcano on White Island in the Pacific Ocean. The views are beautiful and unforgettable. The helicopter also enables the traveller to take in the enormous agricultural space, learn about the structure of land use and estimate the sizes of farms.

Agritourism increasingly involves using the plane as a means of transport. Within only a few hours the plane transports us into another world with completely different conditions. This includes a change of country and language, through different seasons of the year and times of the day, to a different culture, religion and law. On the one hand, a flight by plane is an attractive form of travel, but, on the other hand, it requires considerable concentration. First it is necessary to prepare travel documents, which include passport, visa, vaccinations and ticket. Before departure it is necessary to prepare and pack the luggage, which also requires consideration and some organization. At the airport a passenger needs to remember many things and be alert all the time. Ticket and luggage control, passport and customs control, security control and finally boarding card control. Eventually the passenger boards the plane and takes a seat. The steward gives safety instructions and the flight begins. A short flight, which lasts from 1.5 to 2.5 hours, enables travel among European countries. Flights over the Atlantic take 7–10 hours. Air travel to the Far East or South America lasts 12–16 hours.

03 11 2001 12:22

Fig. 15.6. A tourist helicopter over Lake Whakatipu near Queenstown, New Zealand (photo by M. Sznajder).

The flight from Europe to New Zealand with necessary changes takes even more than 30 hours. During that time, flight attendants will serve a meal and drinks depending on the airline and class we are travelling. The possibility to observe the terrain we are flying over depends on access to the window, the weather and the flight altitude. The flight altitude is 10–11 thousand metres and enables a rather general recognition of the terrain. The lower we fly, the more details we can observe. The view from an altitude of 7000 metres is clear. It is the last 15–20 minutes before the landing, when the plane is flying below 3000 metres, that are especially good for observation. A novice recognizes hardly anything. With time one gains more experience. Then one can clearly see mountains, lakes and seas from the plane windows. When we fly over agricultural land, it is possible to assess the structure of land use, the percentage of arable land and permanent grassland, forest land, crops, buildings and land under water. It is also possible to analyse the type of farming on the basis of land use. On landing we need to be alert again. It is the same course of procedures again: passport control, checking all declarations, baggage claim, frequently phytosanitary and veterinary control and customs control; sometimes there is a dog walking among the passengers looking for food or drugs. Finally, there is the question as to how to reach the place where we are going to sleep if there is nobody waiting for us. Because of the time difference, which is sometimes several hours, many people suffer from the disorder called jet lag. This disorder results from the time shift and reveals itself in the form of alertness at night and sleepiness during the day. The symptoms may continue even as long as 2 weeks after the flight. In favourable weather conditions, when we have a seat by the window, we can change a long flight into an agritourist flight. On such a flight we can observe the fantastically changing agritourist space and recall many facts from economic geography. Box 15.3 presents a description of agritourist observation during a flight around the world, whose route is marked in Fig. 13.1.

Box 15.3. Agritourism on a flight around the world

When flying long hours we can variously pass the time: impatiently waiting long hours for the end of the flight, watching films, reading literature or simply sleeping. A trip around the world may have the character of an agritourist adventure. The journey described here is in fact a combination of a few, which altogether make up a trip around the world. After all, a continuous journey would last 50–55 hours. The route of the trip goes from Poznań via Warsaw to Frankfurt and then via Los Angeles to Auckland. The return route goes via Singapore, Frankfurt and back to Poznań by Eurolot.

Early in the morning we leave Poznań for Warsaw. We leave Lake Kierskie on the right and head for Warsaw. Below districts of Poznań can be seen and a series of trough lakes near Kórnik and Zaniemyśl. We soar high very fast. Sleepy towns and villages, hardly recognizable fields, the landscape covered in clouds does not give good views. Shortly before the landing in Warsaw fish ponds in Raszyn can be seen, a railway line leading to the south of Poland, Ursynów and the plane taxis on the runway. We quickly change for the plane to Frankfurt. Now for 2 hours we shall be watching the agritourist space typical of Europe. A flight over Europe seems banal and obvious to the Europeans; however, later we shall see that it is unique. The green colour is dominant – the lighter green of arable lands and the darker of forests. The arable lands are usually cut into irregular quadrilaterals, divided by roads, motorways and rivers. Scattered small towns and villages can be seen. The closer to Frankfurt, the more undulating the terrain. Finally the last stage of the flight, the city centre, the River Main, the cathedral and the plane taxis along the runway again. In Frankfurt we change for a Boeing 747. The monitors show the flight route. This really helps us to observe and recognize the area. We fly north towards Jutland. Unfortunately, clouds cover the view. But on the coast of the North Sea they clear away and perfect visibility accompanies us right to Los Angeles. On the Danish coast, on the North Sea, white spots can be seen – it is a concentration of wind turbines for generating electricity that are arranged in rows to form a square. How many are there? Between the wind turbines there is a small ship sailing, perhaps for conservation. The location for the wind turbines is well chosen; it ensures the effectiveness of these investments. The plane heads northwest. We leave Iceland on the left far behind. The two or three hours of flying over the North Sea pass quickly. Then the plane flies far beyond the Arctic Circle. The clouds form the shape of snow stripes driven northwards by the wind. The plane heads for central Greenland. The clouds clear away and we can admire magnificent views again, the northern part of the Atlantic Ocean, the light blue sea. We are approaching Greenland. This can be noticed because of the icebergs floating on the ocean, which from this altitude look like small white spots, then powerful rocky islands with icebergs floating here and there and finally the huge fiord Scoresbysund (Ittoqqortoormiit). The sea cuts deep into the land. High naked rocky mountains with steep slopes falling into the sea, a white snowcap on the plateau of the mountain. A wonderful view of glaciers winding like rivers in a valley of majestic cliffs falling into the waters of the fiord. It is the beginning of winter. The valleys are not covered in snow yet. Then it is only the interior, a white snowy desert with rocky inselbergs, which can be seen from time to time. These must be mountain ranges. Greenland again shows the ends of the east coast in full splendour and for a goodbye reveals a huge iceberg. Baffin Island – this is Canada. What can we see there? Naked rocks, lakes, no trace of life. Sunbeams are reflected by the earth. It is impossible to distinguish between the land and huge pack ice. Then for a few hours we fly over completely frozen northern Canada. This must be the tundra. The land becomes a little green – these must be the tundra meadows. I look forward to seeing the taiga. It should start at any moment; after all, these are huge forests. But they are nowhere to be seen. We fly over Great Slave Lake.

After a few hours of flying over wilderness the first road appears. I am curious as to where this road comes from and where it leads. Probably it cannot be a hardened road, but perhaps it is? At last I can see some people underneath. Civilization starts. To my surprise the tundra ends and there is no taiga; great fields suddenly emerge, in which hard wheat is grown. This wheat is the best to make bread. How different this view is from the view of European agricultural space. Grey is the dominant colour. It can be seen that the fields were ploughed after the last harvest. They usually have the shape of large squares. All the land is frozen. From time to time farm centres can be seen or even villages and small towns. A 2-hour flight over such space corresponds to the distance of nearly 2000 kilometres. It gives an idea what an immense field of corn production the eastern part of Canada is. The grassy terrain begins again, it goes up. We are approaching the Rocky Mountains. Long rocky mountain ranges parallel to our flight and wide valleys between them. Empty infertile land – perhaps only cattle graze here. It is clear that the soil lacks water. There is no green at all. The sandy, brown and black colours are dominant. From time to time we can see that wild currant shrubs cover the mountains. There is no agricultural use of the land visible. The landscape becomes even more unfriendly as we fly further. This is the area of Great Salt Lakes. Far and wide there is white salt instead of the lake, only occasionally some water. How can people live in such conditions? Then the famous Colorado River and the Grand Canyon, which did not look amazing from that altitude. From time to time round green fields. In this wilderness the production of fresh green vegetables is possible only due to sprinkler irrigation. Finally, it is California – unfriendly dry sandy land. We are approaching the giant city of Los Angeles. The streets form quadrilaterals. Hundreds of thousands of houses and huge car traffic. The megacity looks stunning. How can one live here? Suddenly the plane enters clouds. It lands in Los Angeles after 13 hours of flying. The city greets us with an overcast sky, light rain and a temperature of only 12°. After a 6-hour break we take off, heading for Auckland. We take off just before midnight. We fly over the Pacific almost all night. We are flying over a boundless immensity of water. Unfortunately nothing can be seen underneath. We can only imagine Hawaii and numerous Pacific islands below. The plane lurches slightly from time to time. The weather is turbulent. The Pacific is not smooth. We shall be crossing the Tropic of Cancer, the equator and the Tropic of Capricorn. The morning on the next day means we have lost one day in life. This always happens when we fly from America to Asia or Australia. The dawn greets us still over the Pacific. Far and wide, white creases can be seen – they must be the rough ocean. The plane goes down. We are approaching Auckland. At last, the outline of beautiful green New Zealand can be seen: the green mountainous terrain of the Coromandel Peninsula with herds of sheep. We land. It is windy, rainy and cold. We have flown half the way. Having visited this beautiful country we set off further west to Poland. After a visit to this beautiful country we are going further east, to Poland. First to Singapore. After a 3-hour flight we reach Australia. First Moreton Island, next we can see the Brisbane metropolitan area. Some forested areas – this must be the Great Dividing Range and next the interior of Australia. We can see clusters of trees and dry land covered with grass (these are probably grazing grounds for beef cattle). Straight roads, occasionally you can see symptoms of civilization – homesteads of cattle farmers. We are flying along the shore of Carpentaria Bay. At times we can see small islands. The day is drawing to a close. The plane is still flying north. We just have a glimpse of the mountainous Timor, covered by a tropical forest. We are landing in Singapore after dark. The airport is beautiful, you can see that it is the gateway to Asia and the World of Islam. We are landing in Europe at night. The night is going to be long, as we are flying from the east to the west. It is a pity that we will not be able to see these agricultural landscapes. The route is leading us over Sumatra, the Inidian Ocean, just off the fertile Ceylon, next the Arabian Sea, the sandy Oman, boundless mountains of Iran and Turkey, the Black sea and Romania. It is dawn. Over Hungary we can see land. We are flying south of Budapest. We can see narrow and long fields at the Danube. Now it is clear to us that this is Europe, but the agricultural space here is also beautiful and unique.

The Agri-guide

It is helpful to have an adequate understanding of agricultural and natural phenomena and the agritourist space. People look at it, but often they do not understand it. When visiting a particular area many tourists value a professional explanation; otherwise a study trip is not as valuable as it could be. Sometimes it happens that an excursion group travels along the same route on two different coaches. There is a guide on one coach but not on the other. We may notice that when the coaches stop at the arranged points, the tourists travelling with the guide have a better understanding of the goals of the trip, whereas the others do not realize what they are visiting. Understanding agritourist space requires the assistance of a professional guide, who might even be called an interpreter of agritourist space. Not everyone is able to give a competent and interesting commentary on the areas being visited. This demands that the guide should have a breadth and depth of knowledge concerning geography, history, ethnography, culture, buildings, animal and crop production, horticulture and ecology.

The interpretation must be attractive, communicated with appropriate words and interesting stories. Simultaneously, agriculture and its agritourist space must be presented in an honest way and in accordance with scientific accuracy.

Guiding groups of tourists is a certain kind of acting, but it can also be compared to the role of the shop assistant. If the person selling an agritourist commodity cannot in fact say anything specific about it, they lose a chance to get new clients (Fig. 15.7).

Weekend Agritourism

City dwellers who own cars are more and more interested in weekend recreation. At the same time, farms have a sufficient potential to manage the increasing demand for weekend agritourism. As one may guess from its name, the idea is to use the free days at the end of the week for recreation, from Friday afternoon to Sunday. In addition, several times a year there is a chance for a so-called long weekend, which lasts up to 5 or 6 days. On the one hand, weekend tourism is a pastime offer for city inhabitants and, on the other hand, it causes better use of the resources of the farms specializing in agritourism. Weekend tourism enables an agritourist farm to use its resources not only during the high holiday season in summer but also during the other months of the year. Good road connections from the city to the country affect the development of weekend tourism. We shall describe weekend tourism, taking the island of Texel in The Netherlands as an example during a rather unattractive season of the year – the cold and wet end of January (Box 15.4).

Fig. 15.7. The president of a company manufacturing Roquefort sheep cheese in the caves of the Massif Central near Rodez, France, personally presents the company and its technological products (photo by M. Sznajder).

Box 15.4   Weekend tourism on Texel

Weekend tourist trips to Texel are an attractive opportunity for many Europeans. The island of Texel belongs to the West Frisian Islands archipelago situated in the North Sea, parallel to the coast of The Netherlands. Texel is the largest island of the archipelago. There are several villages here with permanent residents, where agricultural production takes place, and the main town of Den Burg. The landscape is flat, typical of The Netherlands, with an abundance of green areas. On a January afternoon the whole island is covered in snow. Texel is known around the world for the breeding of a specific sheep breed, which is also named Texel. The sheep are stockily built with wide heads. This breed is widely used for cross-breeding. Texel sheep are an inseparable element of the island's landscape. The possibility of easy travel to the place of stay and easy return travel is an important part of weekend tourism. On a winter January Friday in Den Helder a long queue of cars wait to embark on a ferry to Texel. The travel is the bottleneck of weekend tourism. Likewise, on the island itself there are jams when disembarking from the ferry: a long line of cars, which slowly drive along the only road to their accommodation. There are a lot of beautiful farms and small hotels on the island. In the central place, Den Burg, restaurants and souvenir shops wait for weekend tourists. On Sunday evening tourists take a ferry to the mainland again and return to cities.

The Agritourist Trip

Group trips are an important part of the agritourist business. Frequently they have the character of a study trip, whose goal is to learn about agriculture and living in the country in the areas visited. A group trip is successful if it is well planned and then put into practice. When preparing a trip it is above all necessary to set its goal, the route and the places to visit, the people to meet during the trip, the places to sleep and have meals in and the means of transport the group will be travelling by. When organizing an agritourist trip it is necessary to precisely indicate the date and time of departure, the way of financing and the costs, including the tickets, insurance, vaccinations, protective clothing, food, etc. The timing of the trip is particularly important. A well-chosen time guarantees achieving the set objectives. In former times, in the 19th and 20th centuries, May picnics were common in Europe, because in May nature is in full bloom. Nowadays trips are made almost all the year round. After all, agritourists can be shown many interesting attractions related to agricultural production, agritourist space and the country at any season of the year. Certainly, November and December are not very interesting in northern Europe. It is difficult to attract tourists in that period. In the south of Europe agritourist trips can be made almost all the year round.

When organizing an agritourist trip it is important to set the fees, costs and ways of financing them in advance. It is not a good situation if it turns out that the costs have been miscalculated and it is necessary to pay extra costs. Some organizers of trips set a financial trap offering the trip at attractive low prices, deliberately concealing some costs in order to claim extra charges during the trip, thus putting the tourists in an awkward position. For example, this pertains to parking fees, breakfasts at hotels, admission charges, etc. The trap of offering a cheap trip that in fact costs much more causes annoyance and tension in the group. Ideally, tourists should clearly know before the departure what they are paying for and what possible extra charges they have to bear. Sometimes one of the more difficult elements of a group trip is to control escalating emotions. People experience various emotions for different reasons. The ability of the tour guide to ease tension is necessary. A tour guide who can manage people's emotions is extremely valuable. Experience shows that there are many typical situations that cause rising emotions. As early as the beginning of the journey, when taking seats on the coach, emotions rise. Everybody wants to choose the best seat for themselves. However, frequently tourists cannot easily make a decision as to which seat is the best. Hunger, thirst and fatigue are also reasons for different tensions. Hungry people are easily annoyed. Therefore, it is necessary to plan mealtimes carefully and plan the time well for relaxation and tourism. Emotions escalate when a large group of people want to check in quickly at a hotel but the reception staff are inefficient. Improper account settlement may also spark a conflict. Sometimes there are even conflicts concerning the route of the journey if it is not planned and accepted in advance. The trip programme sometimes happens to be overloaded and the participants do not have enough free time for themselves. Such a programme should be altered promptly and the least attractive points should be left out. Sometimes conflicts arise when the whole group goes shopping. The trip participants have different opinions concerning the products worth buying. As a rule, the participants should purchase all goods individually. Before entering a shop or supermarket we set the venue point and the amount of time we can devote to shopping. Besides the methods that help to avoid or ease tensions, there are also methods that help to build a good atmosphere in the group. Providing praise, thanks, an occasional joke, an interesting story, a song and a time to relax are key guide skills. The guide's personal charisma is also very significant.

Examples of Study Trips

Agricultural conferences are usually connected with agritourism trips. The International Dairy Federation (IDF) organizes such trips during every congress and conference. So also do the other international and national associations, e.g. the Polish Association of Agricultural and Agribusiness Economists organizes annually a 1-day agritourist study trip for its members. Because there are about 200 agricultural economists from all over the country participating, there are three or four routes offered, which are identified on the trip programme. The economists have a chance to familiarize themselves with the problems of agriculture or food processing or the development of rural areas in the part of the country they are visiting. Below we give a short description of a 1-day agritourist study trip to the Lake Wigry National Park. The Lake Wigry National Park, situated in north-eastern Poland, area 14,800 ha, comprises Lake Wigry and the adjacent part of the Augustów Forest. The goal of the trip was to familiarize the participants with the functioning of the Lake Wigry National Park, the tourist and agritourist attractions offered in the park and the regional food products and, finally, to provide some relaxation for the participants. The whole group was divided into four subgroups setting out from the village of Wigry and finishing the trip with a group regional meal in Krzywe, where the head office of the Park is located. Each subgroup had the same programme components, but they were arranged in a different sequence. This facilitated the use of the people and equipment involved in the operation of the agritourist trip and efficient tourist service. The trip programme included a 30–40-minute ride in horse-drawn carts around the forest (the carts took about 20 passengers), a half-an-hour walk through the forest in the direction of the landing stage and a trip on Lake Wigry in the boat Pope John Paul II travelled in. During the cruise there was consumption of smoked lavaret and European whitefish, the fish characteristic of this lake.

In addition, a 40-minute ride on the forest train was planned. During the train ride there was a chance to offer tourists the local cake – *sękacz litewski* (Lithuanian pyramidal cake). The next point of the programme was a visit to the Lake Wigry National Park Museum and a walk along the educational path. Eventually there was a meal served by a local family catering company specializing in regional dishes – *kartacze*. The mealtime was pleasantly accompanied by a family band. The above example shows that for the organization of such a study trip a company serving tourist traffic is critically important. The Polish Association of Agricultural and Agribusiness Economists performed the role. This guarantees not only coordination of the programme but also the inflow of tourists into agritourist areas.

Opportunities for study trips to learn about agritourism exist even in countries where this form of tourism is not spoken about yet. An agritourist trip to Zanzibar, a land where spices are grown, is described in Box 15.5.

Box 15.5. Agritourism in Zanzibar

Unguya, i.e. Zanzibar, has long been known to people (Krapf, 2003). Bantu tribes settled there about 4000 years ago. Then the island was occupied by the Egyptians, Phoenicians, Hindus, Persians and Arabs, who all treated this place as a transfer point in the slave, ivory, gold and precious wood trade. In the 15th century Portugal put an end to the golden age. It ruled for as long as 300 years. At the beginning of the 19th century the Sultan of Oman became interested in Zanzibar and, having taken the island away from the Portuguese, he moved the seat of the sultanate's capital there. Nevertheless, the place could rather be named 'Spice Island'. This is how it is referred to in many guidebooks. If somebody wants to become familiar with the spice plants growing on the island, they should take part in one of many spice tours offered in every corner of the island. A tour is not expensive, local guides are excellent (sooner or later they will guide the tourists to their uncle or brother's plantation) and at the end of the tour a drink of spice tea or spice coffee is de rigueur. What these drinks contain least is tea or coffee. However, it is not necessary to take part in organized tours to see and fully understand the abundance of nature on the island. It is enough to travel by coach a few kilometres out of the city and the nose becomes the best guide. By the side of the sun-scorched road there lie jute sheets with grains of black or green pepper and chilli pods drying, and right next to them the sweet-smelling cloud of clove or vanilla fragrance is suspended in the air. One can stop at such an improvised drying place with the intention of taking a closer look at the flower buds of a clove tree. The island also surprises with a tremendous number of flowers and fruit. Papayas, pomegranates, pineapples, passion fruit, lychees, mangoes – the list is very long. Huge, nearly 30 kg, jackfruits seem to lie almost ownerless on the verge of the road. There is nothing to worry about – their smell could betray any enthusiast for someone else's property. One can travel around the island on a cheap coach, on a bicycle rented at an affordable price or in a car. This can be easily arranged at any hotel or at the 15th-century entrance gate near Jamituri Park. In view of the fact that Zanzibar is an island and it is rather difficult to escape from, there are very few formalities (Krapf, 2003).

Planning an Agritourist Trip

A trip to an agritourist farm for several hours may be spontaneous, especially if it is situated relatively close to one's place of residence. In this case it is not necessary to make special plans for the trip. It is enough to locate the place on a Global Positioning System (GPS) map and one can set off. A several-day agritourist trip is usually a complicated undertaking, which requires detailed planning and making necessary reservations for means of transport and places to sleep, in particular. For the time being an agritourist trip is still an individual enterprise both for a travel agency and for a particular person. An individual agritourist trip may be regarded as a unique original project. The process of planning a trip and making necessary reservations is very time-consuming. Contemporary sources of information and methods of communication, electronic reservation, the possibility to make payments by credit cards and e-banking enable very precise planning of the trip. The more carefully we plan the trip and the better we fix all formalities, the more time we can spend realizing the planned goals, and thus the trip will be more fruitful. Planning and preparing a trip is made up of several tasks. We shall analyse these with the example of an agritourist trip to India, which one of the authors of this book made in March 2008:

- Specifying goals of the trip;
- Gathering information related to the trip;
- Specifying towns to visit;
- Planning places, institutions and companies to visit;
- Planning travel routes and places to sleep;
- Trip budget;
- Establishing contacts with people who will facilitate the agritourist trip;

- Marking points of the journey on a GPS map;
- Entering all necessary numbers into the memory of a mobile phone;
- Safety on the journey;
- Reserving tickets and places to sleep;
- Trip documentation.

When planning an agritourist trip, access to Wikipedia, You Tube, the websites of towns, companies and institutions, Google Earth and other Internet resources provides us with priceless service. Wikipedia provides encyclopaedic information and You Tube videos from the places to visit. Google Earth gives us a satellite map of the Earth, which is often very detailed. With Google Earth we can locate places to visit and the way to get there to an accuracy of several metres. Google Earth also provides a graphic representation of the climate of a place. These electronic sources enable us to prepare a plan of the trip realistically enough for us to have an impression that we know them very well when we arrive there.

Case Studies

1. Large farms are capable of arranging an internal network of cycling paths. Learn about the spatial structure of a large farm. Plan internal agritourist cycling paths for it. Plan the tourist's places to stay, showing agricultural and natural places of interest.

2. For agritourists travelling longer distances in a country by coach or train prepare a commentary related to plant and animal production observed through the window.

3. Predicting the role of an agri-guide prepare examples of interesting stories related to farming that future tourists will find interesting.

4. Make a detailed plan of an agritourist trip to a country on another continent:

- use Google Earth to make a detailed plan of places to visit;
- use Internet resources to learn about the places to visit;
- mark the location of places to visit in a GPS receiver;
- plan places to sleep and means of transport;
- plan a health security programme;
- try to establish contacts with people at the places to visit;
- plan the budget of an agritourist trip.

16 Agritourist Farms and Enterprises Around the World

This chapter briefly presents 22 agritourist farms and enterprises from different parts of the world, namely Belgium, France, Germany, Italy, New Zealand, Poland, Russia, UK, USA and Zanzibar. The authors of this book visited all these places except two. Due to the passage of time the current condition of these farms may be slightly different from the description.

The goal is to present their diversity, attractiveness and general principles of management. In view of the limited character of this publication only brief information concerning the organization and management of the enterprises is presented.

Conventional Farm

Many agricultural specialists and regular tourists enjoy visiting ordinary farms. These specialists are farmers or other workers related to the agricultural sector. The incentive for the visit is interest in the latest production technologies and a willingness to compare one's own farm with farms owned by other farmers in other countries. In principle each production farm can be considered as a potential agrotourism farm and can be visited. Visiting a production farm involves a certain risk for the person receiving guests. This particularly applies to the possibility of transferring diseases. However, immigration and veterinary services take care to eliminate the risk of transferring diseases. Farmers allowing a visit to their farms are honoured that it is their farm that has been chosen. They expect to hear the visitors' opinions about their management standard. Therefore, visiting ordinary production farms brings benefit both to the receiving farmer and to the visitors. It is during such 'host' visits that the idea of agritourism is fully realized. However, the number of such visits is relatively small and it is not the mainstream trend of agrotourism.

Two examples of milk-producing farms that sometimes allow visitors into their premises are described below. Both examples come from New Zealand: the first is Mr and Mrs Kennedy's farm in the North Island and the second the farm of a Dutch immigrant in the South Island. Examples of ordinary farms that sometimes receive guests could be provided from any corner of the world.

Dairy farm in the North Island

New Zealand is a world leader in milk production. Production is orientated towards cost minimization and the export of dairy products. Therefore, the production system is completely different from that in Europe. Pastures are the basis of farmers' welfare; therefore they are used very carefully. The contemporary economy moves sheep to less profitable, more difficult, mountainous, poorer pastures. Better pastures are used for dairy farming. When travelling around New Zealand we can see wide areas of pasture divided into paddocks in which the cattle cannot be seen at all, and individual paddocks where large herds of animals are pasturing at the moment. The Kennedys' farm, situated south of Auckland, farms 1200 cows. All animals are kept under the open sky all the year round. The only buildings for livestock are milking parlours. The farm has a modern carousel milking parlour and the milking process can be watched from the gallery near the parlour. An interesting picture, which is worth being shown to tourists, is the moment when cows in a long line queue to be milked or come back after milking. The visiting time for non-specialists should not be longer than 30–45 minutes. Specialists can spend 3–4 hours, using this time mainly to exchange experience.

Dairy farm in the South Island

Similar to the previous farm this farm does not have an agritourist character. However, it is an interesting case, which could be discussed with farmers from many countries, in view of their unwillingness to sell land to foreigners. A Dutch emigrant bought agricultural wasteland, which he converted into a beautiful dairy farm. He divided the whole area into paddocks, on the borders of which he planted trees or he enclosed them with wire. He built a milking parlour, a building for farm staff and an irrigation system. Water is pumped from a depth of 150 m into the irrigators. In spite of the fact that the area is characterized by a relatively high rainfall, the farmer waters it all the time. This enables high levels of pasture production. He keeps the cows under the open sky. The farmer is involved in seasonal production. All cows calve in September within a period of 2 weeks and then a 10-month period of milk production follows. This system is synchronized with the dynamics of grass growth. During the 2 winter months, July and August, cows are not milked at all. Visitors admire the strict production regime, the courage to invest in agricultural wasteland, the irrigation system, which may seem irrational to visitors due to the high rainfall, and the perfect 'architecture' of the paddocks for grazing animals.

The Biggest Ranch in the World

The world's biggest farm, Kings Ranch in Texas, allows tourists to visit it and presents a rich tourist programme to the visitors (Kings Ranch, accessed April 2008). Kings Ranch was established in 1853 by Captain Richard King. His dream was to found a big farm. King's farm was an example of appropriate farm management. Today it is perceived as the birthplace of the American ranch industry. The ranch plays an important role as the leader in the world agricultural business. It is a vertically integrated company starting with cattle breeding and finishing with the packaging of processed meat products. It is claimed to be the biggest farm in the world. The ranch activity area amounts to 825 thousand acres. The cattle are sold both for consumption and for breeding purposes. Besides the business activity the ranch is also open to visitors. A special centre for visitors was established, where they are offered various attractions, starting with guided agricultural tours and finishing with encounters with nature. A shop and museum were opened as an additional source of income. The Kings Ranch Museum houses a collection of saddles from all over the world, guns used by cowboys during the period of the ranch's existence, replicas of historic Texan flags, historic wagons that were used to conquer the Wild West and vintage cars. This example shows that combining agricultural production and agritourist activity on a large scale not only is possible but can also be economically justified.

Agritourist Farms for Younger Children

Little children value direct contact with animals. Animals satisfy a lot of their emotional needs. The contact with animals must be controlled. On the one hand, children are very frightened of big adult animals, which may scare them. On the other hand, they may be cruel to small animals. Children should have contact with healthy animals that cannot infect them with animal diseases. This section presents three examples: first, Pennywell Farm in the United Kingdom; second, Goat Court in northern Poland; and third, Marshalls Animal Park in New Zealand. Each of the farms attracts the same sector of clients – children with parents. Only Goat Court offers the chance to stay on the farm for a few days. The two others are orientated towards tourists who want to spend a few hours on the farm.

Pennywell Farm and Wildlife Centre in Buckfastleigh, Devon

'All roads lead to Pennywell' was written in the advertising brochure of the farm, which is undoubtedly a tourist attraction of Devonshire, located between the towns of Exeter and Plymouth. The farm advertises itself as 'your favourite farm – attractive in any weather', and the owner adds: 'I buy your worries and I sell joy.' The farm can be reached by car (there is a car park for several dozen cars) or by coach (special discounts for children). If you take care of the natural environment, i.e. if you arrive, for example, by bicycle, you get special discounts. The farm invites visitors 7 days a week from March to October, and in winter every weekend, during the school winter holidays and at Christmas. The offer of the farm is not only to watch the life on it and encounter animals but also a whole lot of other attractions: picnics, educational paths, obstacle courses (Fig. 16.1), children's theatres, pony rides, games and plays, practical activities and many, many more events suited to the children's age and interest. The farm itself, situated in very picturesque and scenically varied surroundings, occupies an area of 45 acres (18 ha) and there are about 400 animals on it. The owner started the business with a loan of £60 thousand. At the beginning about 13.5 thousand guests a year visited him. In 1998 there were as many as 50 thousand visitors and the farm was chosen the best attraction of the region for all seasons of the year. In 2006 Pennywell Farm and Wildlife Centre for the fifth time beat off stiff competition from other farm attractions to scoop the 2006 Farm Attraction of the Year, in its category of 20,000 to 50,000 visitors per year, organized by National Farm Attractions Network. The judges said that:

> The number of farm attractions in the UK has increased enormously and the competition is fierce … but for the past 16 years, visitors have enjoyed a consistently safe, good value, hands on, informative farm experience and Pennywell Farm easily merited being winner of their category because of the high quality of visitor care.
> (High quality of visitor care clinches national award for Pennywell, 6 October 2006, http://www.connectingsw.net.)

In Pennywell tourists spend their money on admission, souvenirs and refreshments. There is also a cafe and a souvenir shop, which also sells food for animals. There are a few people working on the farm, mainly women, who take

Fig. 16.1. Pennywell Farm, United Kingdom, an agritourist farm for children: the obstacle course – wildlife commando course (photo by M. Sznajder).

turns to do all the jobs, which range from looking after animals to serving the guests and working in the kitchen. Everybody is full of enthusiasm and really loves what they do. 'If you don't love your job, you'd better give it up' – this is another life motto of the farm owner. They regard anybody who takes people's time as their competitors. Being full of new ideas, year on year they improve their activity, attracting more and more visitors. Their biggest concern is to ensure tourist flow. However, in spite of the fact that the region is not highly populated, owing to appropriate advertising and the unquestionable attractiveness of the location, the farm successfully attracts people even from remote corners of the United Kingdom.

Agritourist farm Kozi Dwór (Goat Court)

The agritourist farm named Kozi Dwór (Goat Court) is about 20 kilometres west of Olsztyn in the village of Gietrzwałd. The very location of the farm guarantees a continuous inflow of tourists as it lies near the thoroughfare going through a well-known pilgrimage centre devoted to the cult of the Virgin Mary, where numerous pilgrims arrive. The agritourist farm was established on the basis of an old farm specializing in breeding pigs. Kozi Dwór caters mainly for parents with little children, i.e. families who want to spend a few weeks on holiday there. Activities for children are organized so parents can relax. The farm has succeeded in gaining a large group of loyal clients, who arrive at the same time year in, year out. There is a windmill on the farm and a pond with a boat, but it is goats that are the main attraction. The children can gain various skills in looking after animals, which have been arranged especially for them. Also, sharing meals and spending free time together are a big attraction for the families and people staying here.

Marshalls Animal Park

This is a farm zoo near Tauranga on the North Island of New Zealand (see Fig. 16.2). Some farms abandon specialized animal production and transform into a farm zoo, where tourists are offered various attractions. Marshalls Animal Park is an example of a farm that abandoned sheep production, transforming the diversified terrain into an animal park, which tourists can visit. The owners did not set a goal for themselves to gather the maximum variety of animal species in a relatively small area of about 30 ha. They provide relatively big, enclosed pastures for a few species. Tourists are free to enter some runs, but they are not allowed into others where more dangerous animals are kept, e.g. yaks. The goal is to provide maximum contact with animals for tourists. In Marshalls Animal Park the following species of animals are bred: goats, various breeds of sheep and cattle, exotic breeds of pig, yaks, various species of poultry, etc. Tourists can visit the park on foot or they can take a ride. Besides animal watching and contact with them, tourists are also offered a picnic ground, a playground, animal feeding, pony rides, water with trout and eels and many other attractions. The farm earns

Fig. 16.2. Marshalls Animal Park: a general view (photo by M. Sznajder).

income from admission charges, animal feed sold to tourists and the organization of picnics and parties. The owners think that the return from 1 ha used for the animal park should equate the profit gained from its best alternative use. To generate the same net income as 1 ha of kiwi fruit about 200 tourists must visit the farm each year. To generate the same net income as 1 ha of dairy farm about 300 tourists must visit the farm each year.

Agritourist Farms Breeding Wild Animals

When travelling in the country we have a chance to visit farms breeding undomesticated wild animals. There are many species of still undomesticated animals bred for farming purposes all over the world, e.g. ostriches, fallow deer, deer, crocodiles, snakes, crayfish, fish, etc. Some of the farms are open to visitors whereas others do not offer this possibility. In Poland there are some very interesting entities dealing with the breeding of wild animals 'at large', such as the Polish konik farm in Popielno owned by the Polish Academy of Sciences (Polska Akademia Nauk, PAN) and the European bison farm in the Białowieża National Park. In Spain and France bulls are bred to be used in bullfights. In spite of the fact that they are domesticated animals, the example of such a farm is included at this point because of the bull's belligerent nature. Farmers in many countries organize special commercial parks where wild animals are bred as an agritourist attraction. This chapter describes a Safari Park in Świerkocin, Poland, a deer farm on the agritourist farm Deer Pine Lodge near Rotorua in New Zealand and the keeping of wild forest bees, which are rare nowadays, on a bee farm owned by the Shulgan Tash National Park in Bashkiria.

Safari Park in Świerkocin

This is located in the Lubuskie province and is a large enclosed area of 120 hectares where wild animals can be watched from car windows (see Fig. 10.5). The animal stock includes wild boars, deer, fallow deer, European bison, roe deer and elk. A question arises: what is the difference between a safari park and an ordinary zoo? The main difference consists in the fact that safari parks are visited by car. Animals run freely in them. They also eagerly come up to cars. The whole area is divided into several sectors. In each there are animals from another continent. Having visited the park the tourist leaves the car in the car park and enters another sector. It is possible to buy some animal feed there and feed the smaller animals running freely in the enclosure or to have a tasty meal. Children can have fun in the playground.

Bull breeding for bullfights

In the south of France, as in Spain, bullfights are organized. A bullfight is a fight involving a man and a bull. The objective is to kill the bull 'artistically'. Bullfights arouse a lot of passion. There are both staunch supporters and opponents of bullfights. In France one of the places where bullfights are organized is the Coliseum in Nîmes. Bullfights require bull breeding. Bull breeding for bullfights is a source of income for farmers in southern France, but it is also a tourist attraction. One has to pay an admission charge to visit a farm breeding bulls for bullfights. Bulls for bullfights are partly bred on farms in the Rhône delta, and partly, during autumn and winter, in the Central Alps, where, having a lot of feed and space, they mature and grow large. In spring they are transported to farms in the Rhône delta.

Deer Pine Lodge

Deer farms, which cannot be found anywhere else in the world, are characteristic of New Zealand. Deer are bred for venison, which is exported chiefly to Germany. The antlers (velvet) are ground and sold as an aphrodisiac in East Asian countries. Deer breeding in New Zealand started in the 1970s. Wild animals were caught in forests and used for farm breeding. In New Zealand more than 1.8 million deer are farmed. Usually deer farms are not orientated towards agritourism. However, the farm Deer Pine Lodge near Rotorua combines agritourism of the 'farm stay' type with deer breeding and forestry. Visitors to the farm notice the high fences. Lower fences are not possible, because the deer could jump over them easily. The deer are very timid and one must be careful lest a herd of stampeding deer should trample on tourists. The farm stay programme includes a 2–3-hour sightseeing tour of the deer farm and learning about the production of antlers, which Fig. 16.3 presents.

Fig. 16.3.   Deer Pine Lodge: frozen deer antlers (velvet) (photo by M. Sznajder).

Forest apiary in Shulgan Tash

A visit to the Shulgan Tash National Park has an agri-ethnographic character. Besides the Bashkir lifestyle one can also learn about a forest apiary in a remote village. A forest apiary is an extensive and old-fashioned method of bee-keeping. The forest apiary described is situated in the South Urals, in the Shulgan Tash reserve. The place can be reached only by an off-road car. The journey is long and full of tension – mountain tracks going up and down. A tree beehive is simply an artificial hole scooped out in a tree, which is inhabited by bees. More or less one hole out of every four that have been prepared is inhabited by bees.

The forest apiary occupies a large area of forest, where there are 400 logs. A tree used for bee-keeping is characterized by the fact that, besides the hole scooped out in its trunk, the top of the tree is cut. The tree is used as a beehive for 200–300 years. Bee-keepers take care of the bees four times a year. The bee-keeper has to display exceptional courage. A real forest bee-keeper is one who will dance the bee-keeper's dance on the cut treetop. Forest bee-keepers (Figs 16.4 and 16.5) spend the production period in a desolate area, looking after bees and protecting them from bears. They have a lot of work. They visit every tree beehive at least four times a year. The bee-keeper climbs up a tree and removes the lath, which is nailed to the tree. This safeguard prevents bears from taking the honey. Here and there beams are suspended, which are supposed to disturb the bear. The bear trying to steal honey is hit by the beam hanging on a rope. In this situation it is unable to steal the honey. Forest bee-keepers claim that it is difficult to see a bear in a forest, because it is a timid animal, but they say that these animals often observe people from hiding. If we do not cross their way, we

Fig. 16.4. A family of forest bee-keepers (photo by M. Sznajder).

Fig. 16.5. Forest bee-keepers and their equipment (photo by M. Sznajder).

are not in danger. At the moment the forest apiary in Shulgan Tash is a potential agritourist facility, because few tourists reach its neighbourhood.

Gardens and Orchard Farms

For many years botanical gardens and palm houses have been popular with city dwellers. Both in large Paris and in relatively small Belfast, gardens were established and hothouses constructed to satisfy people's interest in seeing specimens of exotic plants without the need to travel. With time, parks and gardens developed both in small, poor and forgotten places and in the vicinity of royal parks, like, for example, the gardens of Versailles (Fig. 16.6) or the hanging gardens in Mumbai. Another trend is city inhabitants' compelling wish to stay in orchards. The goal of such stays is to learn about the production process and, especially, to participate in picking fruit.

Given below are four different examples of these agritourism enterprises. Only the first of the four is a typical farm. The second example is the garden of Heligan, which was established and then forgotten. The third example is a huge palm-house project – a hothouse financed by the European Union and called Eden. The last of the examples is a natural park in a valley formed in consequence of an eruption of Tarawera volcano, North Island, New Zealand. The development of rural tourism based on gardens and orchards, the examples of which were the described British and New Zealand enterprises, (see fig. 16.7) is one of the ways to stimulate the development of rural areas and their multifunctional use. As can be easily noticed, it is creative people – their ingenuity, competence and skills – that play the most important role in the stimulation of

Fig. 16.6.   The gardens in Versailles, France (photo by M. Sznajder).

this type of enterprise. These people create concrete and daring projects used in tourism or whole packages of services, thus providing employment for themselves and other people.

Van Vlaanderen: strawberry production in a hothouse in Belgium

Modern strawberry production can be used for agritourist purposes. Modern technologies have actually enabled prolonging the harvest season over the larger part of the year. The idleness period is very short, comprising only 3 months – November, December and January. The Van Vlaanderen garden farm in Belgium produces strawberries in a hothouse. The host, though not orientated to agritourism, allows specialists and those interested in visiting the premises to enter free of charge. Visiting a farm producing strawberries according to this technology is interesting to tourists. Strawberries ripen at the height of a standing man, which makes picking them easier. The soil is completely covered with plastic film to prevent the growth of weeds.

The Lost Gardens of Heligan

These gardens in Cornwall remained forgotten for over 70 years and now they are the largest garden area under restoration in Europe. They were established

Fig. 16.7. The garden in Hamilton, New Zealand (photo by M. Sznajder).

in Victorian times, but, at least from the end of the Second World War, when the last gardeners working in them died, they remained idle. In spring 1991 the gardens were still under a layer of fallen leaves, tangled ivy and fallen trees, but a year later a group of people were working on their restoration and preparing them to be open to visitors. The gardens consist of:

- The Northern Gardens, which comprise a whole gallery of gardens in various styles, including the Italian garden, the New Zealand garden, the northern garden with arbours, the garden with a sundial and the vegetable and fruit garden (peaches, grapes, citrus fruits, melons, pineapples in hothouses, herbs and flowers).
- The Jungle and the Lost Valley, with subtropical vegetation and a view of a fishing port. The following plants can be found: palms, bamboos, rhododendrons and numerous exotic trees. The area is diversified by ponds, bridges, secret passages and paths.
- The Secret Garden.

Enormous outlays coming from various sources were used to restore the gardens, but today they attract thousands of tourists. The enterprise gains income from admission charges, the sale of flowers and pot shrubs and the sale of souvenirs.

The Eden Project

Another example is the Eden Project near the town of St Austell in Cornwall, established in a 60-metre-deep china clay pit occupying 15 ha. The construction of two huge greenhouses called biomes started in 1994 and in 2001 they were opened to tourists. One of them, called the Humid Tropics Biome, occupies an area of 1.55 ha: it is 55 m in height, 200 m in length and 100 m in width. The other biome occupies an area of 0.65 ha: it is 35 m in height, 100 m in length and 65 m in width. Figure 16.8 presents a picture of the project.

Coconut palm trees and rice, cotton and hemp, bamboos and oranges, olives and grapes and hundreds of colourful flowers 'grow before the eyes of visitors'. In the gardens of Eden they can admire a wide range of vegetation from different climatic zones without the need to leave the country. The bigger of the two biomes houses the humid tropics flora, the smaller one reproduces the warm climate for the vegetation of the Mediterranean, southern Africa and the southwest of the United States. The remaining area of 12 ha outside the greenhouses is used for the vegetation of the local climate. The mission of the Eden gardens is the promotion of understanding and responsible management of relations between plants, people and environmental resources to keep a balance between them. In the garden there are about 100 thousand plants, representing 5000 species of most climatic zones of the world.

The first constructions of the Eden Project were finished in 2000. The total cost was to be about £77 million, i.e. over 460 million zlotys. In May 1997 it was subsidized by £37 million from the European Union (the Millennium Commission). In addition, it was sponsored by the European Regional Development

Fig. 16.8. The Eden Project (photo by Tamsyn Williams, reproduced with permission).

Fund, the National Lottery, local organizations and autonomies, banks and commercial companies. According to the business plan presented, this unusual garden and research centre were supposed to attract at least 750 thousand visitors a year, whereas during the first year of its operation 1.91 million visitors arrived. The enterprise employs 600 people, who were unemployed before.

How the World Began

Waimangu Volcanic Valley near Rotorua in New Zealand was named *How the World Began*. It is a place of geothermal activity. The area is covered by jungle and adapted for tourists. The educational objective is to show the world at the moment when life began. On 10 June 1886 at 2.00 there was a powerful eruption of Tarawera Volcano near Rotorua. The petrified inhabitants were awakened by the enormous noise, which could be heard at a distance of 500 km. The mountain, illuminated by thousands of lights, ejected masses of ash, which fell in the vicinity, causing considerable changes in the topography of the area. The volcano is different from others. It is an 8-kilometre crevasse stretching over a range of peaks, from which lava squirted. The eruption of the volcano caused 20 serious changes in the topography of the area.

As a consequence of the eruption a valley was formed, later covered by the jungle, which witnessed the return to life after the volcanic destruction. There are many geothermal phenomena there, including the world's largest lake of boiling water, Frying Pan Lake, where the average temperature is about 45–55°C, but at the deepest point reaches 200°C (Fig. 16.9).

Fig. 16.9. Waimangu Volcanic Valley near Rotorua in New Zealand: the world's largest lake of boiling water – Frying Pan Lake (photo by M. Sznajder).

Vacation Farms and Farm Stays

Vacation farms are the oldest form of agritourist activity. They were promoted as early as before the First World War. There were known summer holiday resorts where holidaymakers from the city arrived for a rest. In the 1970s, when it turned out that the Workers' Holiday Fund was not efficient enough to serve holiday-makers, 'a holiday under the pear tree' was popularized. Nowadays the main trend of agritourism derives from those times. Farmers offer tourists beds and sometimes meals. The tourists organize their free time individually. Thus, most farms providing agritourist services can be called vacation farms. Their main disadvantage consists in the fact that holiday services are offered only for a few months in the year and after that period they are normal farms. However, increasingly, extra services are offered to holidaymakers.

A short stay on a farm that lasts 2 or 3 days is called 'farm stay'. It will be described with an example of a farm in New Zealand. This country is visited by numerous tourists from Europe, America and Asia. They hire cars and by means of them they reach the most remote corners of the country. They often choose to spend the night on agritourist farms. These farms vary greatly. Te Hau Station Farmstay, near Gisborne, is an example. The farm, whose area occupies more than 5.7 thousand acres, is situated in the central-eastern part of North Island. The farmer breeds sheep and his wife deals with agritourism. Before arriving, it is necessary to book a room. Each guest means special preparations for the hostess. She has to go to the nearest shop, which is about 40 km away. The hostess awaits guests in the afternoon. She serves them dinner, which is the main meal. The hostess welcomes the guests simultaneously, trying to make a nice and friendly atmosphere. The guests take a while to unpack their belongings and have a wash. Then, for a while they chat about different things with the host, who is responsible for keeping the conversation going and he can do it well. About 11 pm the host definitely needs to go to bed. Tomorrow he has a long working day. The guests also go to bed – after all, they have had a long and tiring day of travelling.

In the morning cooked breakfast is usually served. The hostess suggests a tour of the farm. It is possible to feed the animals, some of which are quite exotic, give milk to the calves and then drive a car round the farm. The farm is huge with a diversified terrain, including many hills and mountains. The tour of the farm takes about 4 hours. On returning from the tour tourists who are staying longer are offered a meal and 1-day tourists are expected to sign the guest book. Then the hostess asks a typical question: 'Would you like some tea or coffee?'. This means that it is time to pay for the stay and leave the farm.

Cheese Tourism

Cheese is one of the most delightful farm products, and that one fits in with agritourist activity very well. There is an opulent range of cheeses (Fig. 16.10). In France their number is estimated at 1000 types. Most frequently farmers expect tourists to show an interest in a specific kind of cheese and how it is manufactured and finally they expect tourists to purchase it. Goat, sheep and buffalo cheeses are offered as special attractions. Being agritourist products, cheeses can be well combined with other products, especially wine. This section will present examples of agritourist farms from the United Kingdom, Germany, Italy and France. There are descriptions of various combinations and organizations of farms manufacturing goat, sheep and buffalo cheese and those producing ecological cheese in mountain areas. Cheese production is also combined with running a restaurant or keeping a shop, where cheese is the dominant commodity. Usually farms offering cheese do not expect tourists to stay on the farm for very long. The maximum time does not exceed 2–3 hours. The agritourist programme is typical: familiarizing tourists with the manufacturing process and tasting and purchasing cheese. In addition to this, the farmer may expect tourists to buy and consume a meal at the restaurant near the farm. As an additional attraction, the cheese-processing plant is enriched with decorations depicting some elements of the production process that the visitor cannot watch personally, e.g. a dummy shepherd surrounded by sheep. The farms and companies that are more orientated towards agritourism organize short presentations, which include an interesting story and even a musical accompaniment. Below we describe five examples of agritourist farms that use cheese in their activity. They are Lynher Farms and Dairies, where Cornish Yarg cheese is manufactured (United Kingdom), Hohensteiner Farm in the Swabian Mountains (Germany), where ecological farm cheese is produced, a monks' farm (Italy, near Rome), where cheese is manufactured from buffalo milk, Pelardon Farm in Languedoc (France) breeding goats and manufacturing goat cheese, and Société Roquefort producing sheep cheese in the caves of the Massif Central (France).

Fig. 16.10. Cheeses developed by an agritourist farm near Rome in Italy (photo by M. Sznajder).

Lynher Farms and Dairies

Cornwall is a region of beautiful coastal landscapes and unusual environment, both natural and created by humans. On an area of 500 acres (over 200 ha) of land leased from the Prince of Wales, between Bodmin Moor and the valley of the River Lynher, is the farm called Lynher Farms and Dairies, which manufactures atypical cheese wrapped in nettle leaves. Until recently the owners bred 250 cows in two herds and about 250 young cattle with a small area of corn plantation for feed. The farm is served by a team of five people. Twelve people work in cheese production, the café, which also serves cheese, employs two people and there are up to 26 seasonal workers, whose work includes picking nettle leaves.

The farm was subsidized by European Union structural adjustment funds. It was granted the means for diversification of the agricultural sector, development of tourism, natural environment conservation and development and development of human resources, including new employment opportunities, which all amounted to about £0.5 million.

Currently the farm produces about 100 tonnes of Cornish Yarg cheese (blue maturing cheese wrapped in nettle leaves) and fresh soft cheese from about 0.75 million litres of milk, which comes only from their own farm. Cornish Yarg cheese is manufactured in accordance with a special formula that is kept secret by the family; this is considered to be exclusive cheese, bought only for special occasions. Also it is quite expensive, mainly because of the manual work that goes into its production. The farm sells the cheese mainly to large networks of shops, such as Marks & Spencer or large department stores, such as Harrods. In summer the owners open the farm to visitors. On average the number of visitors arriving reaches about 8000 individuals and groups a year, and each tourist leaves about £4.50 on the farm.

Hohensteiner Farm

Robkas and Albkas are ecological cheeses from Hohensteiner Farm, situated in the Swabian mountains in south-western Germany, about 100 km south of Stuttgart. This farm simultaneously has an agritourist, ecological and mountain character. The farm produces cow's milk in accordance with ecological standards. Cheese is made from the milk. The farmer has used various structural funds from the European Union to invest in his enterprise, which consists of the following components:

- The farm, where cattle feeds meeting the requirements of ecological production are produced; the farm also produces ecological milk;
- The processing of ecological milk into cheese, including ecological cheese;
- Agritourist activity orientated towards the presentation of the production process and then selling the cheese to tourists. In addition to this, the farm shop offers food products manufactured by the local people.

The visiting time on this farm is not long for a group, the maximum being 1–2 hours. During this time the tourist can watch the cheese manufacturing process, purchase cheese, consume it on the premises or take it home. The farm is orientated towards tourists passing the farm on the way, staying only for a day. The farmer does not expect the same tourist to visit his farm again. The owner estimates the number of visitors at 200 groups a year, which makes 15–20 thousand people. Tourists do not have to pay for admission into the farm. The owner instead expects income from the sale of cheese, which is much more expensive than in an ordinary shop. Apart from the cheese the farmer sells products manufactured by the local people. This is an additional offer extending the sale range. When we are in the shop section we can observe the cheese production process behind glass. Two kinds of cheese are manufactured there. Additionally there is a small conference room that can be used. The interior decoration also looks interesting. According to the farmers' declarations the cost of establishing such a farm reached about 2.5 million Deutschmarks.

Buffalo Farm

South of Rome there is a multidirectional agritourist farm run by monks of the Catholic Church. The milk buffaloes are the attraction of the farm

(Fig. 16.11). The buffaloes have acclimatized themselves well in the south of Italy.

In the countries of the European Union there are milk production limits called milk production quotas. Buffalo milk is not subject to these limitations; it can be produced without limit, because it is not a component of the milk quota. The enterprise consists of three parts. The main element is the farm, which, besides plant production, maintains buffaloes. The farm has a shop that sells products manufactured on the farm or by the neighbours, and a restaurant where local products can be tasted, including buffalo milk cheese. The tourists who have never tried buffalo milk will eagerly come there again and try it as part of another adventure. The farm is orientated towards the tourist passing on the way. It is run by monks and nuns. The tourist has a chance to watch the buffalo stock and observe the buffalo cheese manufacturing process behind glass. The buffaloes themselves are an attraction in Europe. Buffaloes are divided into two types: riverine and swamp buffaloes. Riverine buffaloes are unsuitable for milk production. The average yearly milk yield is 1000 litres and very good stock can yield even as much as 4000 litres a year. Some people are afraid of buffaloes thinking they are dangerous animals. In fact, wild buffaloes are very dangerous animals. Milk buffaloes, contrary to common expectations, are very placid animals.

Fig. 16.11. A farm in Italy manufacturing buffalo cheese: buffaloes in the run (photo by M. Sznajder).

Pelardon

This farm is situated in the south-eastern part of the Massif Central in France. It is an example of an agritourist farm with diversified offerings. The building is typical of southern France, built of stone, not very conspicuous outside, but decent-looking inside. The farm logo is simple – a yellow goat's head with the inscription 'Fromages' (Fig. 16.12), which means goat cheese is made here. Both the stock and processing are in one building. Tourists have an opportunity to see goats, which are very interesting animals. Humans communicate with them easily. Pelardon cheeses resemble in taste maturing cheeses of the Brie type. At the end of the stay tourists have the opportunity to taste some cheese or consume a full meal. The farmer expects only weekend tourists and only then does he open the restaurant (Fig 16.13).

At the restaurant it is possible to taste goat cheeses, with which wine is served. The farmer is orientated towards weekend tourists and therefore his restaurant is open only three days a week, i.e. on Friday, Saturday and Sunday.

Fig. 16.12. Pelardon agritourist farm in France: the logo inviting tourists to the farm (photo by M. Sznajder).

Fig. 16.13.   Pelardon agritourist farm in France: the restaurant receives guests only at the weekend (photo by M. Sznajder).

Agritourist Restaurants

Restaurants are becoming an integral element of modern agritourism. Most traditional agritourist farms offer home meals. Agritourist restaurants differ from traditional restaurants in that they are owned by a farmer, who offers local dishes, which are usually made from the food produced on his farm. Agritourist restaurants can also be a means of promotion for food products of agricultural processing companies. Well-served meals made from home products may encourage tourists to buy them. Sometimes the decoration of a city restaurant refers to farming or, more frequently, rural motifs, which can be seen in Fig. 16.14. This section gives examples of agritourist restaurants or ordinary restaurants referring to rural or agricultural traditions, including the Sodas restaurant offering dishes of the national Lithuanian cuisine.

Sodas

Sodas farm, situated in Trakiszki, is an example of the tradition of the Lithuanian minority in Poland being used for agritourist purposes. A large group of people classified as the Lithuanian minority live in the district. Sodas farm, which is situated in an ethnically Lithuanian area of 11 ha, used to grow cereals and breed beef cattle and pigs. The income gained was not sufficient. The farmer decided

Fig. 16.14. Using the elements of old manufacturing technologies for decoration of a restaurant interior, Rennes, France (photo by M. Sznajder).

to switch to agritourist business, subordinating all his farm to the activity. The quality that differentiates the farm from other agritourist farms is the Lithuanian character. In Puńsk the farm opened the Sodas restaurant with national Lithuanian dishes such as: *čenakai* (stewed vegetables with minced meat and spices), *vedarai* (sausage stuffed with mashed potatoes, bacon and spices), *blynai* (Lithuanian potato cakes stuffed with meat), *cepelinai* (cylindrical potato dumplings stuffed with meat, onions and spices), *kibinai* (ravioli) in beetroot soup, etc. At the restaurant one can also buy national cakes such as *šakotis* (pyramidal cakes) and *skruzdelynas* (pastries topped with honey and poppy seeds). The range of national Lithuanian products is likely to be extended. Moreover, the farm has been transformed into an agritourist centre for families who intend to spend free time in a Lithuanian environment. The Sodas generates income from the following sources: the restaurant with the Lithuanian menu, the sale of regional Lithuanian products and receiving agritourists. The economic effectiveness is increased as a result of providing home food products for the restaurant.

Dairy in a City

A great attraction of Mumbai is the Aarey Milk Colony, which covers a total area of 3166 acres of land. The Aarey Milk Colony is a unique complex of structures producing milk. It is situated in the northern part of the Indian metropolis of

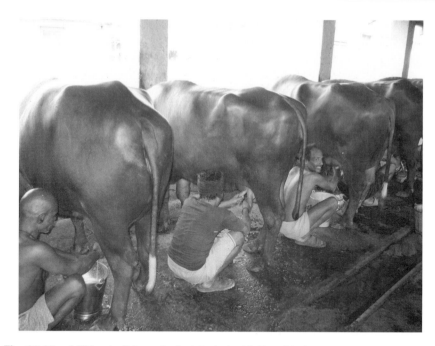

Fig. 16.15.   Milking buffaloes, India (photo by M. Sznajder).

Mumbai, in the beautiful Borivaldi Park. The colony provides fresh milk for the local residents. Individual farms are leased to individual lessees by the government of the state of Maharashtra. However, the government has control of them. The system of keeping, feeding and milking buffaloes is interesting. One may also see dung being dried for fuel. The colony is one of the most popular tourist attractions in Mumbai and draws a large number of visitors from all over the country. Apart from the dairy experts and cows, the Aarey Milk Colony comprises an observation pavilion, picnic facilities, milk plants, gardens, a nursery and lakes. The observation pavilion is situated on a hill near the entrance to the Aarey colony, from where one can view a captivating landscape. The main feature of the Aarey Milk Colony is the fine gardens, which spread out over almost 4000 acres of parkland. The Aarey Milk Colony has an average of 16,000 cattle reared on 32 farms. Figure 16.15 shows cows being milked in the colony.

Kumis Therapy

Agriculture provides products and services that are used in the process of the treatment and rehabilitation of people. These processes, carried out with the means available to agritourism, are more and more frequently included in agritourism. There is a wide range of possibilities within this direction of activity, but most of them are still waiting to be discovered. This section describes kumis therapy in the health resort or sanatorium in Yumatovo near Ufa. In the near future kumis therapy may become one of the mechanisms stimulating agritourism in Bashkiria. At the moment there are more than ten kumis health resorts in this republic. Kumis is a product made from mare's milk. It is a national drink of many people of Central Asia. In olden times the inhabitants of the immense steppes of this region of the world used kumis to quench their thirst during the hot summer. Kumis has been known since at least 500 BC. Herodotus mentioned it. Prepared mare's milk undergoes double fermentation: milk fermentation and alcohol fermentation. To make kumis milk is a slow process (6 hours) and involves mixing appropriate cultures of bacteria. Then for a period of 3–21 days the milk undergoes the maturing process. Kumis is unsuitable for long-distance transport. It has such an effervescent quality that it may break the bottles in which it is transported. Kumis has a refreshing taste but it may also have an intoxicating effect. Many curative qualities are attributed to kumis. It has importance in the treatment of lung diseases, especially open tuberculosis, alimentary tract diseases and nervous diseases.

Combining horses, kumis production and therapy in a health resort is an interesting business strategy. In one enterprise the whole cycle is organized: production, processing and use of the product. Yumatovo health resort near Ufa consists of the health resort buildings, a kumis plant and a farm breeding milk horses. In spite of the fact that the health resort is orientated mainly towards Russian patients, soon it is likely that agri-therapy will also be offered to foreign patients.

The livestock farm is another facility (Fig. 16.16). Horses graze the steppe. They are milked five times a day. In one milking, 200 g of milk is obtained from

Fig. 16.16. Yumatovo kumis health resort: horse herdsmen (photo by M. Sznajder).

Table 16.1. The characteristics and income strategies of the agritourist farms and entities described in Chapter 16.

No.	Name of farm	Location	Type of enterprise	Agritourist products and services offered	Characteristics of products and services consumers	Sources of financing
1	Dairy farm	North Island, New Zealand	Milk production farm	Presentation of modern technologies and production organization	Farmers, milk manufacturers mainly from New Zealand, but also occasionally from all over the world	Income from production Income from the milker training centre
2	Dairy farm	South Island, New Zealand	Milk production farm	Occasionally a tour of the farm	Farmers, milk manufacturers and specialists mainly from New Zealand, but also occasionally from all over the world	Income from production
3	Kings Ranch	Texas, USA	Stock ranch	Guided tours of the historical and agricultural farm Trips out into nature Saddle shop	Organized groups of tourists and individuals	Income from animal production Income from processing Income from agritourism Income from the sale of souvenirs
4	Pennywell Farm and Wildlife Centre	Buckfastleigh, Devon, United Kingdom	Agritourist farm	Children's contact with animals Animal feeding Obstacle course 'Toy farm' playground	Mothers and whole families with little children	Admission charges Income from the restaurant
5	Kozi Dwór (Goat Court)	Gietrzwałd, Warmińsko-mazurskie province, Poland	Agritourist farm	Longer rest among animals	Reservation. Families with little children	Accommodation charges
6	Marshalls Animal Park	North Island, New Zealand	Zoo farm	Picnic area, domestic animals	Reservation. Families with little children	Admission charges Selling feed for animals

7	Świerkocin Safari Park	Świerkocin, Lubuskie province, Poland	Association	Watching animals from car windows Feeding animals Cafe Playground	Motoring tourists	Admission charges Income from consumption
8	A farm breeding bulls for bullfights	The Rhône delta, Central Alps, France	Farm	Watching bull stock possible	Any tourist	Income from the bull stock Admission charge
9	Deer Pine Lodge	Ngongotaha, near Rotorua, New Zealand	Agritourist farm	Farm stay Touring the farm	Foreigners, reservation	Income from beds Income from consumption Income from deer production
10	Forest apiary	Shulgan Tash, Bashkiria	National park farm	Excursion into the Ural Mountains Watching the collecting of honey	Occasional tourist	Income from the sale of honey Costs of transport must be covered
11	Van Vlaanderen	Ghent, Flanders, Belgium	Garden farm	Watching the manufacturing and distribution process possible	Occasionally specialists	Income from sale Income from strawberry production
12	The Lost Gardens of Heligan	Pentewan, St Austell, Cornwall, United Kingdom	Part of a former agricultural estate	Visiting a unique park possible Purchasing pot plants possible	Anybody interested	Admission charges Sponsors' support Sale of plants Souvenirs (e.g. DVD)
13	The Eden Project	St Austell, Cornwall, United Kingdom	Company	Visiting the wet biome and the hot biome possible 5000 species of plants from all over the world	Anybody interested	Admission charges Sponsors' support Cafes and restaurants Souvenirs

(Continued)

Table 16.1. Continued

No.	Name of farm	Location	Type of enterprise	Agritourist products and services offered	Characteristics of products and services consumers	Sources of financing
14	Waimangu Volcanic Valley	Rotorua, the geothermal ecosystem at the site of the valley formed in consequence of the eruption of Tarawera Volcano, New Zealand	Company	Looking at geothermal phenomena and plant development possible	Domestic tourists, but above all foreign tourists	Admission charges Sale of a wide range of souvenirs
15	Te Hau Station Farmstay	North Island, New Zealand	Sheep and agritourist farm	Farm stay, a tour of the sheep farm and participation in the farm work, home meals, tours of the surrounding area	Both domestic and foreign agritourists, also groups of specialists	Income from sheep production, bed and board provided for guests, the sale of products from wool
16	Lynher Dairies Cheese Company Ltd, Pengreep Dairy	Ponsanooth, Truro, West Cornwall, United Kingdom	Milk farm and cheese factory (company)	Observation of processing Cafe with possibility to taste cheese	Anybody interested	Income from the sale of cheese to hotels Income from the sale of cheese to tourists Cafe
17	Hohensteiner Farm	The Swabian mountains south of Stuttgart, Germany	Ecological agritourist farm and cheese factory	Observation of the cheese manufacturing process Shop with local farming	Anybody interested	Ecological milk and cheese production The EU subsidies for the unfavoured areas Sale of ecological cheese to networks Sale of ecological cheese to tourists

18	Buffalo farm	South-east of Rome, Italy	Monks' church farm Processing Restaurant	Contact with animals possible Tasting buffalo cheese possible Watching cheese production possible Meals	Anybody interested	Plant and animal production – wholesale Plant and animal production – direct sale Restaurant
19	Pelardon	Central Alps, France	Goat farm	Watching the goat stock possible Processing milk into cheese Restaurant	Weekend tourists who want to have a meal	Income from the sale of goat cheese to networks of shops Income from the restaurant, including goat cheese dishes and wine made by local manufacturers
20	Sodas Restaurant and Guest Rooms	Trakiszki and Puńsk, Warmińskomazurskie province, Poland	Agritourist farm	Recreation on the farm Lithuanian restaurant	Holidaymakers Tourists looking for meals with a certain national tradition	Income from the agri-hotel business Income from folk cuisine
21	Aarey Milk Colony	Mumbai, India	Buffalo farm	Observation of production process	Individual tourists who are interested in buffalo milk production	Income from milk production
22	Kumis therapy	Health resort, Yumatovo, near Ufa, Bashkiria	State health resort	Koumiss therapy Familiarizing oneself with the kumis production	Health resort patients from Russia	State budget Small contribution from the patients
23	Spice tour[a]	Zanzibar, Tanzania	Farm producing spices	Familiarizing oneself with the spice production process Tasting	Foreigners	Admission charge

[a]See Box 15.5.

a mare. A good mare gives a maximum of 1 litre of milk a day, which makes about 300 litres a year. In some months there is overproduction of horse milk, so powdered horse milk is manufactured from the surplus. In the periods when there is a shortage of horse milk, especially in winter, kumis is made from the reconstituted horse milk. The taste of kumis depends on the place where it is manufactured. In the Yumatovo health resort its smell and flavour encouraged one to drink. Another kumis, made in the Ural Mountains, had the strong smell of mare's milk.

Conclusions

The description of agritourist farms and entities presented in this chapter enables us to make a few generalizations concerning their operations. In spite of the large market for agritourist products and services, the farms and entities providing agritourist services have to concentrate on attracting clients. In fact, a spontaneous inflow of tourists is only occasional. In order to achieve this goal, agritourist farms and entities apply various strategies concerning the offer of products and services (portfolio), price strategies and, finally, promotion strategies. Table 16.1 summarizes the agritourist farms and entities described in this chapter. The description includes the name of the farm, its location, the type of enterprise, the agritourist products and services provided (production portfolio), specification of the client sector for which the farm or entity has prepared the offer and the source of financing.

Relatively few of the entities presented specialize exclusively in agritourism. Agritourism is only part of the activity of the agritourist farms or entities. Usually they combine tourism with other forms of activity, and sometimes it is a sort of hobby.

Considering the range of the services and products, it is possible to say that a diversity strategy dominates. This means that farms usually offer their clients many different products and services. As far as clients are concerned, most agritourist farms are not orientated towards a specific segment of consumers of the products and services they offer. Only a few farms are orientated towards a specific segment, especially the ones focussed on little children and those focused on farm stays. It is a common strategy to address the offer of products and services to all potential tourists. This means that the income from agritourism is uncertain or small.

Agritourist farms apply various price strategies in order to ensure income. Some farms do not gain income from agritourist activity at all. This applies to production farms, which occasionally receive experts on agricultural production. Such farms feel honoured by the fact that other farmers visit them, and an exchange of experience takes place. Other farms obtain payment for agritourist services. However, it also turns out that in most cases the income from agritourism is only part of the income of those farms.

Three main techniques of obtaining payment for the services provided can be observed. The first and second apply to tourists who do not stay overnight. Tourists observe the production process presented in an interesting way for free.

After the end of the sightseeing, the farmer offers the product for sale. The structure of this technique is as follows: free entry into the facility, free observation of the production process, tasting (usually with a charge), paid-for dinner, the relevant component of which is the observed product, purchasing the product 'as a souvenir' for home. The second technique of charging is more decisive. The tourist has to pay the admission charge, which includes a certain package of products and services, and an extra charge is made for ordering an extra service. The third technique applies to overnight tourists and holidaymakers. The tourist purchases the service and pays for it several days in advance. This usually applies to accommodation and board.

Case Studies

1. If you live in a big city, explore the recognized and the potential agritourist facilities. Plan a 1-day agritourist trip to those places.
2. Write a flyer about the places for the excursionists to visit.

17 Agritourism on the Edge

The following examples of agritourist activity taken from many countries around the world show that agritourism is blossoming. It offers more and more products and services, some of which cannot easily be classified. There are phenomena related to rural areas that in fact cannot be classified as agritourism, but which cannot be ignored by agritourism, because they are important sources of income for the inhabitants of rural areas. As time passes, agritourism breaks away from the traditional, slightly folk parochialism and becomes a modern business activity. It becomes a subject of interest for external investors. We can constantly see its development and the new opportunities that it gives. Changes in the concept of agritourism go incredibly far. Where agritourism contacts other forms of business activity, there is an interesting zone of interaction. This applies to virtual agritourism, agritourism in big metropolises and using agritourism as a means of promoting farm products and food industry products. The production of alcoholic beverages has always been a big tourist attraction. Universities, agricultural schools, scientific stations and research institutes are willingly visited by people involved in proper agritourism. An interesting example is New Zealand sheep farms, which by a strange coincidence have became an exceptional tourist attraction. Similarly, pilgrimage traffic often complements agritourist activity and contributes to the wealth of the local rural population.

The examples described in this chapter indicate the continiung and wider inclusion of agritourism into modern productive activity in agriculture and business activity in tourism.

Agritourism as a Means of Product Promotion

Agritourism is more and more frequently used as a means of product promotion by the agricultural and food industry. Companies arrange special agritourist facilities such as museums, exhibitions, tasting salons, restaurants or galleries for observation of the manufacturing process. By visiting a food-processing company, tourists become ambassadors propagating the products of the company. Both specialized farms and the food-processing industry use agritourism as a method to promote their products. Below is a description of a soft fruit farm in Koronowo, which receives tourists in the hope that they will spread all over the country the news of the possibility to purchase shrubs. There is also a description of the activity of a Parmesan cheese manufacturer, Latteria Soresina (Italy). There are many more examples of using agritourism for product promotion.

Soft fruit farm

The Licensed Nursery of Berry Shrubs is situated in Buszkowo, near Koronowo (Poland). It grows young strawberry, raspberry and blackberry plants. The company started its activity in cooperation with the Agricultural University of Poznań about 10 years ago. It has established a farm of 400 hectares that deals exclusively with the production of young plants. The farm is well organized in a modern way, with the marketing department working energetically. It applies many methods of promoting their products. Agritourism has also been used as a method of product promotion. The owners welcome groups of visitors and explain the production process in detail. A meal is also included.

Parmesan cheese

In the north of Italy in the River Po valley, near the city of Cremona, situated about 70 km south-east of Milan, Parmesan cheese is made. Real Parmesan cheese is made only in the Po Valley, where the soil contains unique components that enable production of cheese with a unique taste. Parmesan cheese made outside the Po Valley is only a surrogate. The cheese reaches full maturity after being stored for 1.5–2.5 years. The dairies that make Parmesan cheese willingly show the whole manufacturing process. Tourists have an opportunity not only to see how it is made but also, and most importantly, to taste it. The manufacturing process can be observed both in small cheese dairies and in big factories. Latteria Soresina is a medium-size dairy. It buys about 100 million litres of milk a year. The Parmesan cheese manufacturing process is typical and similar to the production of other maturing cheeses. Tourists find cheese maturing rooms the most attractive, where different forms of cheese can be seen. There are several methods of forming cheese; for example, one produces corrugated balls and another makes long cylinders, which weigh about 50 kg. However, most frequently Parmesan cheese is shaped in the form of a flat cylinder. How is such a big ring of cheese cut? It cannot be done with a knife. Special appliances are used for this, which make cutting cheese in half possible (Fig. 17.1). There are also special knives for eating cheese. The dairy allows tourists to visit the premises and taste the cheese.

Fig. 17.1. A demonstration of cutting a cylinder of Parmesan cheese (photo by M. Sznajder).

Agritourism in the Largest Indian Metropolis

The very idea of agritourism in a metropolis of more than ten million people is hard to conceive. The usual associations that go with a metropolis are an enormous built-up area of residential buildings, shops, offices, industry, a labyrinth of motorways and streets and excessive noise. It seems that this is no place for farming, let alone agritourism. However, the idea of agritourism in huge metropolises does not have to be purely abstract (Fig. 17.2). It is enough to realize that every day large agglomerations consume huge amounts of food, which they usually also produce. In many metropolises there are farming, fruit-farming, horticultural or even animal breeding facilities, as well as facilities for the food farming industry and trade in these products, which may constitute a perfect base for agritourism. These facilities have a large agritourist potential, which may be a source of considerable income. It seems that many residents of large cities usually have no idea where food comes from and how it is produced. They also value contact with farming, breeding, food processing and nature near their place of residence. It is relatively easy to satisfy these needs. Hence, a long time ago many cities opened parks, botanical gardens, palm houses and zoos to people. Food industry plants, e.g. breweries, provide the possibility to visit their facilities. The problem is that both the owners of these very interesting agritourist facilities and the residents of metropolises themselves are often unaware of the importance of their potential. Agritourist attractions of big city agglomerations often remain unknown and unexplored.

The urban complex of Mumbai (formerly Bombay), situated on the west coast of India, on the Arabian Sea, has a population of over 20 million and has

Fig. 17.2. Domestic animal sculptures in big cities remind city people of idyllic villages (photo by L. Przezbórska).

a very interesting agritourist potential. Mumbai itself, the capital of Maharashtra State, is full of hustle and bustle and contrasts, especially between the rich and the poor. The city has a considerable number of people living in slums or even without homes. Yet this city offers interesting agritourist attractions – hanging gardens in the south and milk settlements in the north. Besides, it is probably the only metropolis in the world that has a national park within its limits – Borivali National Park. The food-processing industry, especially the dairy industry, is developed in Mumbai. The wholesale fruit and vegetable market is also worth seeing. As in all of India, in Mumbai sacred cows are a considerable attraction – usually these are humped zebus. A great attraction of Mumbai is the Aarey Milk Colony, which is described in Chapter 16.

During one author's stay in Mumbai, he managed to organize a 2-day agri-tourist trip, comprising a visit to two dairies, which included finding out about and tasting Indian dairy products. The Aarey Milk Colony is a farm managed by a Hindu association that provides milk for poor people. The trip also involved observation of the wholesale market for fruit and vegetables and the eye-catching sacred cows.

It is surprising, but an agritourist trip does seem possible in every big city. A question arises as to whether it is attractive enough to compete with the typical tourist attractions of a city. Naturally, if somebody arrives in Mumbai, they want to see the biggest attractions of the city and there are a large number of them. Some agritourist attractions, like the Aarey Milk Colony, are very competitive. Thus, for its inhabitants, agritourist trips in a metropolis may be a viable market segment.

Virtual Agritourism

Not everybody has enough finances or time for agritourism. Sometimes an individual's state of health is a barrier. Virtual tourism and agritourism are becoming a more and more realistic opportunity for millions of people all over the world. Sometimes virtual tourism becomes a form of temporary escape from the problems of everyday life. Virtual tourism may also be an introduction to real tourism. Simply speaking, it is possible to plan and then realize such a programme.

Modern technical facilities like the computer and the Internet and programs like Google Earth, You Tube and Wikipedia are invaluable tools of virtual tourism. It is nice to make a 2- or 3-hour agritourist trip to unknown places. Google Earth provides a detailed satellite image of the Earth. Some parts of our globe are surprisingly clearly visible. This tool gives a chance to see nearly all corners of the world via satellite. Rural areas look very interesting in these images. It is enough to select and zoom in on the image of the part of the world we would like to explore. The satellite image enables us to see the organization of rural areas, buildings, types of roads, the arrangement of fields and larger buildings for animals. Many details are visible. Google Earth users very often upload photographs and descriptions of places. Some sites are linked with entries in Wikipedia. In You Tube archives we may find a video suiting our needs. Modern technical facilities enable the increasing reality of virtual agritourism and provide knowledge and pleasure for many people.

Incidental Agritourist Trip

When travelling to new places or walking around them, we can suddenly find ourselves in an unusual situation, in an unexpected but very interesting environment. We come across an unexpected opportunity for an agritourist trip, so we need to make the most of it. Unfortunately, sometimes tourists look but cannot see, they listen but cannot hear or understand an opportunity. They miss an extraordinary chance.

When walking on the Arabian Sea beach in India towards the fortress in Daman, suddenly the holiday beach changes. One cannot fail to see fishing boats, sheds for drying fish, women busy with work and fishing settlements. By chance, we find ourselves in the centre of fishing and the processing of fish and other seafood. The fishing settlement stretches over a distance of several kilometres. There are movement and work on the beach. On the shallow rocky shore fishing boats are anchored after returning from sea. The catch is carried ashore, where fishermen's wives and daughters sort it. The selection is made on the seashore. Meanwhile children are playing football, bathing or having fun. The fish are carried to primitive sheds to dry in the sun. Along the shore there are many drying sheds made from poles. There are fish hanging on the poles and shrimps drying on cloths. All the waste from processing is carried to the shore and dumped. Plastic cups, paper and dirt can be found all around. The organic waste that remains after fish cleaning is eaten by animals – zebus, dogs and pigs. Fishing settlements are primitive. In the evening villagers play cricket or volleyball.

Dogs curl up and bask in the sun, using the heat of the sunset. It is time to come back. When strolling along the seashore, there is an opportunity to stop at such a village and take advantage of the unexpected agritourist attraction, i.e. observation of life in a fishing village.

Alcoholic Beverages in Agritourism

Tourists are willing to visit agritourist farms that produce alcoholic drinks and liqueurs. Of course, the alcoholic beverages industry is strongly centralized in many countries; nevertheless it allows visitors into their facilities. Some farms serve home-made alcoholic drinks. In fact, for centuries tourists have eagerly visited cellars with vintage wines, breweries and distilleries. However, each country has its own regulations concerning the manufacturing and consumption of alcohol and alcoholic beverages. What is legal in one country may be strictly prohibited in another. In some countries prohibition is in force (usually in Islamic countries), while in others consumption of alcohol in public places is forbidden. In some countries the production of spirits for home use is allowed, whereas in others it is completely forbidden. When travelling the world it is necessary to take due care not to infringe the local law or customs related to the production and consumption of these beverages.

Beer is usually made in industrial processes. This leads to a consolidation of beer brands. The diversity of beer is the richness attracting tourists. Belgium is known for a large number of different types of beer. The Belgians are proud of the fact that they offer more than 300 different kinds of beer of various flavours, for example, raspberry-flavoured beer. Belgium tries to keep this diversity of kinds of beer to attract tourists. In former times a large variety of beers was also manufactured in Poland, but nowadays they are unified, with only a few brands. Some regional beers have disappeared. The observation of the beer manufacturing process is used as a tourist attraction. In Wrocław, in the Spiż restaurant in the Old Market Square, there is a mini-brewery, where beer made on the spot can be tasted. During consumption it is also possible to watch part of the manufacturing process. In spite of the fact that the beer is not of brand quality, it is an attraction for tourists. Special permission is necessary to start the manufacturing and sale of local beer on one's farm.

Wine is a product made from grapes during alcoholic fermentation. The wine world is extremely rich. Vineyards in Europe stretch south from the 49th parallel. Wine is made by the Hungarians, Bulgarians, Greeks, Italians, French, Germans, Austrians, Swiss, Spaniards and Portuguese. The list of countries is long. Various forms of promoting wine to tourists are known. Those who visit southern countries will have seen such attractions as vintage festivals or other customs and traditions. The observation of grapes growing and being harvested, wine production, tasting and visiting wine cellars are a regular element of tourism and agritourism in Hungary, France or Italy. Vintage festivals – bacchanalia (a drunken feast in honour of Bacchus) – are enthusiastically celebrated.

Usually every producer gives their product a unique name. Producers often arrange restaurants and bars in cellars, where one can have a good time. Some producers open their own shops, where they offer various kinds of wine. Tourists are invited to the shops, where they have an opportunity to taste the wine and purchase a selection for home. The way vineyards are managed has a significant importance for the configuration of agritourist space.

Fruit wine production made from wild fruit, e.g. rose hips, is a local Polish tradition. Among agritourist farms there are few that serve fruit wines made from haws or rose hips. This seems to be a niche, which will soon be filled by agritourist farms. Another speciality is the production of liqueurs. They are often produced as home specialities. Some of them have been given commercial brand names such as *śliwowica leska* (plum vodka). However, special concessions and licences are necessary for farms to include liquers in product offers.

Hobbiton – a Stroke of Luck

Some farms become attractive to tourists in spite of the fact that nobody has ever planned such an activity before. We can speak of a stroke of luck. New Zealand is a source of many examples. There may be geothermal activity on a farm, which nobody has previously made accessible. Now it is possible to sell tickets to tourists, who visit this area in masses. Another example is the meadow on the Walim Pass in the Sowie Mountains, which tourists making an adventure trip to Mount Wielka Sowa have taken to. The owner has converted it into a car park and picnic ground.

The owner of a sheep farm near Matamata on North Island, New Zealand, can speak of great luck. It was on this farm and its meadows that scenes of the adventures of Frodo Baggins were shot for the film based on the book by J.R.R. Tolkien, *The Lord of the Rings*. Suddenly the farm became a tourist centre, which provides considerable financial returns. Only initiates know exactly where Hobbiton is situated. While wandering around Matamata, it is difficult to find signposts leading to the farm. The surroundings are extremely beautiful; however, to reach the screened Hobbiton, one has to buy a very expensive ticket at the local tourist information office and go sightseeing with a group. A stroke of luck caused that farm to unintentionally become a large tourist centre.

Pilgrimage Traffic

All over the world, including Europe, there are scattered pilgrimage centres of international importance. They are related to the Catholic or Orthodox Church. Islam also has its pilgrimage centres (Fig. 17.3). The European centres are most frequently connected with the cult of Mary, Mother of Jesus Christ, but also with the cult of angels and saints. In Europe the largest centres of the cult of Mary include Fatima, Lourdes, La Salette, Medjugorje and Częstochowa. Other cult sites are, for example, Assisi, where St Francis is venerated, Mont-Saint-Michel in Brittany, where St Michael the Archangel is venerated, and also the places connected with the lives of saints, e.g. San Giovanni Rotondo, which is related to the life of Padre Pio. Individual countries have pilgrimage centres of local importance.

Fig. 17.3. Pilgrimage places: pilgrimage traffic to Haji Ali's tomb, Mumbai, India (photo by M. Sznajder).

The nearness of a pilgrimage centre enables the local rural community to gain income from providing services for pilgrims. The rural people offer beds to pilgrims, prepare meals, give parking space and sell souvenirs and religious items. In fact, pilgrimage traffic is complementary to agritourist activity. There are many examples of the influence of pilgrimage traffic on the stimulation of activity in small towns and villages. More than 5 million pilgrims visit Lourdes every year. Among them there are a lot of sick people arriving from all over the world with the hope of miraculous recovery. It was there in the Pyrenees that Saint Mary revealed herself to a little girl, Bernadette Soubirous. A spring of water, which pilgrims believe to have healing powers, remains the sign of those revelations. Volunteers help and serve the sick people arriving in Lourdes. All around one can see the local people make a living from the pilgrimage centre. Lourdes without the grotto and spring would remain a small sleepy town in the Pyrenees without much prospect for development.

Mont-Saint-Michel is situated in France on the border of Normandy and Brittany (Fig. 17.4). It is a well-known tourist and pilgrimage destination devoted to the cult of St Michael the Archangel. The history of the place is extraordinary. It shows how the local people flourish and develop when the place is devoted to a cult and how they languish when the cult is forbidden and there are attempts to develop other types of activities. The people living in the surrounding area gain income from serving tourists and pilgrims. Mont-Saint-Michel is a cliff in a zone of extreme tides, probably the biggest in the world. When the tide is low, the cliff is part of the mainland; when the tide is high, it becomes an island.

Fig. 17.4. Pilgrimage places: general view of Mont-Saint-Michel in France (photo by M. Sznajder).

The difference between high and low tide is considerable there and it may even reach 12 m. The information board shows the time when the high tide starts. It is very important information for drivers who park their cars. The cars must be removed from the lower dyke area or the seawater will flood them. There is a small town and a church on the cliff. The legend has it that in the 9th century as a result of a powerful water rise an enormous catastrophic tidal wave rose up, which destroyed all the forests on the coast. Nowadays it is thought to have been an enormous tsunami caused by an earthquake or the impact of a falling meteorite in the North Sea region. Only the cliff remained untouched. The contemporaries thought the island was saved owing to St Michael the Archangel. For a long period of time the cliff was the object of a cult. It provided the local people with the means to make a living. During the French Revolution the building was converted into a prison. For 200 years it was completely neglected. Only about 100 years ago did it again become the site of the religious cult. Tourism developed too. At present it is a tourism pearl of worldwide importance, which millions of tourists from all over the world visit every year.

Agriventure and Agritourism in the Province of Manitoba

Agriculture consultancy in its classic approach, i.e. aimed at spreading rational methods of plant and animal production, is going through a slump in many parts of the world. The developed structures are changing and many people involved in them are looking for new jobs. Programmes of multifunctional development of rural areas are being developed. The province of Manitoba, situated in the Great Plains in central Canada, south of Hudson Bay, has with the assistance of experts developed its original programme named Agriventure. The programme is financed by the local government. Agriventure comprises everything that leads to diversification of business activity in rural areas and everything that adds value to the activity. Currently the programme has three directions of development, namely:

1. Farmer markets (fairs).
2. Agritourism.
3. Diversification of plant and animal production.

Farmer markets are a programme aimed at reactivation of former fairs. On an appointed day of the week, which is usually Saturday, farmers offer their products for sale. It is possible to buy processed frozen meat, fruit and vegetables and handicraft products at the fair. Because Manitoba does not have favourable conditions for fruit production, vegetables and strawberries are usually sold, but there are also processed products, such as preserves, juices and compotes. Handicraft products include sculptures in wood or even soap, flax products, rural-style furniture, artificial flowers, wool, etc.

There are about 200 agritourist farms in the province of Manitoba. They are associated with Manitoba Country Vacation Association (MCVA), which offers two programmes: farms hosting school trips and holiday farms (bed and breakfast). Altogether 22 farms participate in the programme. Many agritourist farms do business outside the association. However, membership in it brings some profit to farmers, especially as far as promotion is concerned (flyers, website: Manitoba Country Vacation Association, accessed April 2008). The cost of membership in the association is 250 Canadian dollars plus Internet costs. Research has proved that farmers earn from 3000 to 5000 Canadian dollars a year on agritourism. In the low season, from late autumn to spring, farms offer different kinds of services from local people, and in summer they host tourists from all over Canada and abroad. One of the most important forms of the business is school programmes. In spite of the wide agritourist offer and extending the period of providing services to all year, farmers cannot treat the agritourist business as the only source of income.

Case Studies

1. Plan an agritourist trip. Use Google Earth to locate an agritourist farm and to find the way to it. Find how much information you can obtain from the program. Check it for links to Wikipedia. Search for the right video on You Tube.
2. The limits of agritourism extend further and further in the contemporary world. Think if there are phenomena verging on agritourism in your country or region. If so, define them. Think if they are used for business. How do farmers and how do tourists benefit from the extension of the limits?

18 The Agritourist Characteristics of Five Countries of the World

Predispositions for the Development of Agritourism

According to the International Programs Center (US Census Bureau, 2008), the total population of the world is currently estimated to be more than 6657 million. The number of tourist trips around the world was estimated to be more than 898 million in 2007 (World Tourism Barometer, 2008). This means that statistically every eighth person travelled, which means that people are extremely mobile. However, similarly to other goods, this mobility is not equally available. This means that a narrow but richer group makes trips, whereas the poor frequently cannot afford their basic life necessities.

According to the facts and figures of the World Tourism Organization the average expenditure of international tourists per one arrival amounted to about $842 in 2005 and in 2006 it rose up to $866. Nevertheless, the market for tourist services is enormous. Table 18.1 shows the most frequently visited countries of the world and the number of arrivals in those countries in millions of people in the years 1995–2006. The greatest number of tourists visited France – 79.1 million arrivals, Spain – 58.5 million, and the USA – 51.5 million. This means that some countries are more attractive to tourists while others are less attractive. Part of this international tourist traffic is related to agritourism. Table 18.1 does not include domestic tourist traffic, but domestic tourists also avail themselves of tourist and agritourist services.

The market for agritourist services is huge. In spite of this, individual entities operating on the market may not feel this enormousness. Each country of the world has its specific predispositions for practising agritourism. Describing these predispositions for all countries of the world surpasses the capacity of this book. Therefore a case-study approach has been applied. This chapter presents five distant countries of the world in respect of their predispositions for developing local and international agritourism. The following qualities should above all be

Table 18.1. Countries of the world most frequently visited by tourists between 1995 and 2006 (the number of arrivals in millions). (Based on the Polish Institute of Tourism; Compedium of Tourism Statistics, 2003; *World Tourism Barometer*, 2003, 2007.)

Countries	1995	1997	1998	1999	2000	2001	2002	2004	2005	2006
France	73.1	66.6	70.1	73.1	77.2	75.2	76.7	75.1	75.9	79.1
Spain	46.8	39.6	43.4	46.8	47.9	50.1	51.7	52.4	55.9	58.5
USA	48.5	47.8	46.4	48.5	50.9	45.5	41.9	46.1	49.2	51.1
China	27.0	23.8	25.1	27.0	31.2	33.2	36.8	41.8	46.8	49.6
Italy	36.5	34.7	34.9	36.5	41.2	39.1	39.8	37.1	36.5	41.1
United Kingdom	23.3	25.5	25.7	25.4	25.2	22.8	23.9	25.7	28.0	30.1
Germany	17.1	15.8	16.5	17.1	19.0	17.9	18.0	20.1	21.5	23.6
Mexico	19.0	19.4	19.4	19.0	20.6	19.8	19.7	20.6	21.9	21.4
Austria	17.5	16.6	17.4	17.5	18.0	18.2	18.6	19.4	20.0	20.3
Russian Federation	–	–	–	–	–	–	–	19.9	19.9	20.2
Turkey	6.9	9.0	9.0	6.9	9.6	10.8	12.8	16.8	20.3	18.9
Canada	19.4	17.7	18.9	19.4	19.6	19.7	20.0	19.1	18.8	18.2
Ukraine	4.2	–	–	–	6.4	–	–	15.6	17.6	–
Malaysia	7.9	6.2	5.6	7.9	10.2	12.8	13.3	15.7	16.4	17.5
Hong Kong (China)	7.8	11.3	10.2	11.3	13.1	13.7	16.6	13.7	14.8	15.8
Poland	18.0	19.5	18.8	18.0	17.4	15.0	14.0	14.3	15.2	15.7
Greece	12.2	10.1	10.9	12.2	13.1	14.0	13.3	13.3	14.3	–
Thailand	8.7	7.3	7.8	8.7	9.6	10.1	10.9	11.7	11.6	13.9
Portugal	11.6	10.2	11.3	11.6	12.1	12.2	11.7	10.6	10.6	11.3
The Netherlands	9.9	7.8	9.3	9.9	10.0	9.5	9.6	9.6	10.0	10.7
Macao (China)	5.1	–	–	–	5.2	–	–	8.3	9.0	10.7
Hungary	2.8	18.7	16.8	14.4	15.6	15.3	15.9	12.2	10.0	9.3

classified as predispositions for the development of agritourist enterprises: unique directions of agricultural production, investment in maintaining the history of farming of the past centuries, development of new farming technologies, well-developed tourist infrastructure and imponderables, i.e. the beauty of the landscape, clean air, historical and cultural heritage. People's wealth, the condition of the technological and transport infrastructure, and a rich farming tradition in rural areas have resulted in very well-developed agritourism and rural tourism in most countries of the European Union, where they are popular mainly among the citizens of those countries. This chapter characterizes five countries of the world with regard to their predispositions for international agritourism: Bashkiria

(Russian Federation), Aoteora (New Zealand), Quebec (Canada), the United Kingdom and Poland. At the very beginning, some general remarks concerning those countries arise. In Russia agritourism is a completely new concept in spite of the fact that many tourist activities can be classified as agritourism. This results from the fact that in practice there are no individual farmers in this country. Thus, there is no group of entities particularly interested in the development of agritourism. Canada is a huge country. However, agritourism is concentrated mainly in the south-western part of the province of Quebec, which is the country's tourist centre. Agritourism is a natural component of the whole tourist system. However, it is mainly Canadians and Americans that are tourists to this centre.

In spite of the fact that New Zealand lies in a very distant part of the world it is an important tourist centre. Agritourism is a significant part of this activity. When travelling around that country the tourist in practice has no chance not to encounter agritourism or not to use it. Agritourist farms in New Zealand are mainly orientated towards receiving foreign tourists. In the United Kingdom a long time ago farms began to welcome tourists, who are mainly inhabitants of the British Isles. Northern Ireland may seem the least advanced in agritourism, though the inhabitants of that country have promoted countless and very interesting agritourist attractions. Poland is the rising star of rural tourism. This is because of what the Polish countryside may offer: tradition, culture, hospitality. We need to stress considerable biodiversity of both agriculture and the environment. Farms are located in very attractive surroundings, among mountains or forests, along rivers, etc. Although rural tourism is still perceived in Poland as tourism for the poor, there is a range of services for more affluent tourists as well.

Bashkiria, the Russian Federation

Hidden agritourist potential

In the Russian Federation agritourism as a tourist idea is still a novelty. Probably it results from the fact that in Russia generally there are no individual farms. Thus, there are no entities that would be economically interested in the promotion of this form of tourism. Agritourism may be developed by other business entities. The mainstream of foreign tourists is bound for Moscow and St Petersburg. Rather few tourists reach other territories. Therefore, it is still the Russians that are the main consumers of tourist services. Some forms of recreation that are offered to people in the Russian Federation can certainly be classified as agritourism or rural tourism in spite of the fact that they do not refer to this idea. Kumis therapy or familiarizing oneself with unique forms of production, like forest bee-keeping, could be regarded as special cases. Russia has powerful potential for the development of agritourism, though at the moment it offers such services mainly to poor Russians at low prices. Doubtless one day in Russia an energetic agritourist sector will emerge, which will present its offer for richer citizens of the world. The Russian Federation is a huge country. Therefore, the agritourist potential in Russia will be presented with the example of the Bashkir Republic.

Geographical context

Bashkiria – officially the Republic of Bashkortostan – defines itself as a sovereign democratic state being part of the Russian Federation. It occupies an area of 146.3 thousand km^2, lying in the central part of the Federation in the southern part of the Ural Mountains and the adjacent lowlands: the Bashkir Pre-Ural and the Bashkir Post-Ural on the border of Europe and Asia. The terrain of the republic is mostly flat lowland stretching in the west, covered in steppes and intersected by rivers. In the east stretches the range of the Southern Urals, reaching 1500 m. Bashkiria has large oil resources. The climate is continental, with a hot summer and a long frosty winter. The capital of Bashkiria is Ufa (population 1.5 million), which is an oil, industrial and university centre. The population of the country is somewhat over 4 million, including 21.9% Bashkirs, 39.3% Russians, 28.4% Tatars and many other nationalities. The Bashkir and Tatar nations belong to the group of Finno-Ugric peoples and most of them profess Islam. The Russians are traditionally Orthodox. A large group of people declare atheism. It is possible to arrive in Ufa by train from Moscow within 26 hours or by plane in 2 hours. At present travelling by car is not recommended.

Islam

When visiting Bashkiria, mosques are commonly seen, with distinctive architecture (Fig. 18.1). Islam is a living religion of those nations, especially in rural areas.

Fig. 18.1. An old mosque in Ufa, Bashkiria (photo by M. Sznajder).

When visiting the small towns and villages in Islamic countries, one needs at least an elementary knowledge of the religion. Currently over 1 billion people all over the world profess Islam in the Sunnite and Shiite denominations or other sects. A pious Muslim has five duties to fulfil: profess the faith (*shahada*), prayer (*salat*), fast (*saum*), alms (*zakat*) and a pilgrimage to Mecca (*hajj*). The mosque is a place of prayer in Islam. Mosques are scattered all around Bashkiria, as churches are in Europe. The people are proud of their faith. Bashkir and Tatar Muslims have a friendly attitude to strangers. When entering a mosque and some homes one must take off one's shoes.

In practice it is possible to go sightseeing around Bashkiria on horseback or in an off-road car. There are few sealed roads. An excursion into the Ural Mountains has an ethnographic, agritourist and natural character. On the way one can visit a *kumisnoe café*, i.e. a place where kumis can be bought (Fig. 18.2). A *kumisnoe café* is an ordinary wooden cottage, where beautiful Bashkir women offer kumis made on the spot, on the farm. The kumis has a very strong horse's smell and taste. The *kumisnoe café* is undoubtedly an agritourist enterprise and simultaneously it is an example of direct sale. One can admire the wooden buildings in villages and small towns. In fact, in all the Southern Urals dwelling houses are made of wood, with characteristic Bashkir architecture. Most roofs are covered in iron rather than thatch, though. Traditional living conditions are another tourist richness. The Bashkirs live in rather primitive conditions, without running water and sewage systems. One can spend some time among the Muslim Bashkirs, who are still rather isolated from the outside world. The Bashkir house

Fig. 18.2. Bashkiria – Southern Urals: a *kumisnoe café* (photo by M. Sznajder).

consists of two parts: the bedroom and the kitchen. The Bashkirs will be pleased to invite a guest to the kitchen, but the living area will remain private.

When visiting the Urals you are certain to see skinny cows looking for food in forests or see them roaming along roads and coming back home for the night. Non-agricultural people who own cows but no land let them graze unattended. Nobody watches them. In the evening the cows will come back to the pen anyway. If not, they are in danger of being mangled by wolves or bears. Finally, at dawn, the Urals reveal the beautiful canyon of the River Belaya, partly enveloped in the morning mist. In spring when the snow melts, the water level rises by as much as 15 m. Canoeing down the river is a big attraction. The Kapova Cave, which is huge, is situated not far away from the canyon. There are paintings on the rocks that were made 20 thousand years ago. They depict various animals, e.g. rhinoceroses and mammoths. Bashkiria is a beautiful country, full of tourist and agritourist attractions, and it is still waiting to be discovered.

Aotearoa / New Zealand

Omnipresent agritourism

The Maori call their country Aotearoa, which means 'the land of the long white cloud'. When sailing across the immense ocean, suddenly a long white cloud can be seen in the distance – it is New Zealand. The culture and tradition of the Maori, the aboriginal inhabitants of New Zealand, and especially their dances, singing and national dress, are a magnet attracting tourists. The Maori have retained their Te Reo Maori language. For an admission fee, one can visit a village where Maori live according to their customs. Because of its climatic conditions, nature and beautiful forms of landscape, New Zealand has particularly favourable conditions for tourism and agritourism. Tourism, agritourism, agriculture and ethnography alternate in every place. The tourist potential is enormous. Almost everywhere, tourists will not only receive the necessary information but also pay the admission charge. When travelling around New Zealand, it is impossible not to encounter agricultural and agritourist attractions, which can be seen in Fig. 18.3. Agritourist farms in New Zealand are primarily orientated towards receiving foreign tourists.

Geographical context

New Zealand is situated on islands in the south-eastern part of the Pacific Ocean about 1600 km south-east of Australia. It occupies an area of 267.5 km^2 and consists of two major islands: North Island and South Island. In addition, there are several minor smaller islands, e.g. Stewart Island in the southernmost part of the country. Tourism and agriculture, especially the dairy industry, make a major contribution to the New Zealand economy. The landscape of New Zealand is very diverse. Most of South Island is occupied by a mountain range – the Southern Alps, with the highest peak Mount Cook (3764 m). The largest glacier is the

Fig. 18.3. Travelling around New Zealand provides many impressions and attractions related to agriculture and traditional settlements. Cows or sheep are sure to block the road (photo by L. Przezbórska).

Tasman Glacier, which is 29 km long. The south-western end of South Island is called Fiordland. Numerous streams flow down the mountains and into not very long but steep-walled fiords, making picturesque waterfalls. The eastern part of the island is dominated by flat lowlands. The mountains of North Island are lower and form a series of ranges stretching parallel to the east coast, separated by wide depressions. The centre of the island is occupied by a volcanic plateau. There are a few huge volcanic cones, three of which are still active. There are also a number of active geysers and hot mineral springs. The coastline is well developed. New Zealand, with the exception of the southernmost end of South Island, has a maritime climate. Summers are warm, with a large number of sunny days, and winters are mild, especially on North Island. The first inhabitants of New Zealand were the Maori, who landed there about 800–1000 years ago. Currently the population of New Zealand is 4 million, comprising 78.6% Europeans, 15.1% Maori, 5.8% other Polynesians and 0.5% others. There are endemic species of plants and animals. Grassland constitutes 95% of the agricultural area, and arable land makes up only 5%.

Quebec, Canada – an Agritourist Centre

Canada can be divided into many regions, but from the point of view of agritourism the southern part of the province of Ontario is one of the most important areas. This relatively small area along the St Lawrence River is a tourist centre.

Agritourism is a natural component of the whole tourist system. Canadians and Americans are the most common tourists there, but occasionally also Europeans.

An agritourist centre is a specific area, usually privileged in respect of tourism, where the tourist and agritourist activity is intensified. Even very privileged areas without historically formed tradition will rarely convert into tourist areas. Usually agritourist and tourist activity in rural areas evolved from the agricultural activity. With time the inhabitants of privileged areas in respect of agritourism noticed that the tourist activity is more profitable than the farming activity. In a relatively small area there are a number of different natural and historic attractions, complemented by the attractions offered by modern society. An agritourist centre is formed in consequence of many farmers' natural abandonment of farming in favour of starting a tourist business. The process takes a longer or shorter period of time. It happens that in some regions farms relatively quickly abandon traditional farming activity, leaving only one or a few farming functions and raising them to the rank of tourist attractions. An agritourist centre will be presented with the example of Côte-de-Beaupré in Canada.

Côte-de-Beaupré

In the southern part of the province of Quebec, there is a tourist centre based chiefly on rural tourism and agritourism. Agritourism plays a significant role in the functioning of this centre. It can be divided into three parts: Côte-de-Beaupré is situated east of Quebec, La Jacques-Certier is in the north-west and the island *Ile d'Orléans* is the third centre. In a way they are three subsystems of a huge tourist and agritourist complex.

Côte-de-Beaupré occupies the area east of the city of Quebec stretching along a distance of about 50 km in length and 5 to 20 km in width parallel to the St Lawrence River. This area is particularly favourable to tourism. The left bank of the St Lawrence River is a strip of exceptional natural and man-made landscape values. There are numerous tributaries of the river, and one can feel the closeness of northern Canadian forests and over 300 years of the history of French colonization. However, during the colonization period, it was mainly an area of agricultural activity. Also, there is energetic business activity in the area and the religious and pilgrimage traffic is flourishing. All this has created a unique area of rural tourism. Quebec City (population 600,000), which is the capital of the province of the same name and is unique for North American conditions, constitutes a tourist base. The very city of Quebec, situated on the St Lawrence River, is a centre of tourism. The river is a great attraction of the region. It connects the Atlantic Ocean with the Great Lakes in America and it is the main transport route for ships sailing from the Ocean into the heart of America. The city is situated on a high scarp, which guarantees a particular attraction. On the other bank, where the island Ile d'Orléans is situated, agritourist activity is also developing. For the citizens of the USA and Canada, Quebec is a substitute for Europe, so they are very pleased to arrive there. Foreign tourists, as well as the inhabitants of the province of Quebec, are the clients of the Côte-de-Beaupré centre.

Apart from the tourist attractions of the region, there are also companies that provide accommodation and food and sell souvenirs. In Côte-de-Beaupré there are as many as 21 farms that offer various agritourist products. The centre of the area is the New France Tract, around which farms are located. During the period when the so-called New France existed, it was the transportation corridor for the state. The New France road runs about 5–6 km off the banks of the St Lawrence River and forms loops at the end. The other key transportation route is a three-lane motorway that enables people to move quickly from Quebec to individual parts of Côte-de-Beaupré. Those two roads are connected by numerous inter-secting roads.

The huge agritourist centre provides employment for a huge mass of people. The agritourist enterprises are usually based on the farmers' old homesteads, which are now the location of businesses that are largely connected with tourism, but which also provide services for non-tourists. The centre functions as hundreds of individual, independent business entities. In principle, these are family busi-nesses. The entities are integrated into a system by effective tourist information and a network of companies supporting tourist services. Sacred places are an important magnet, especially the pilgrimage church of St Anne, St Mary's mother.

Tourists arrive individually in their own cars at independent tourist entities. There are also enterprises organizing group trips to Côte-de-Beaupré for hotel residents. At the reception desk hotel residents can order an excursion along one of the many routes on offer. In the morning a travel agency's coach arrives at the hotel, picks up the tourists and drives them along the agreed route. The coach fare is included in the price, but the tourist needs to buy a ticket each time they want to be admitted to each of the visited places. The tourist is also expected to buy the products offered in the shop or to consume a meal. The effective func-tioning of the tourist information system is an indispensable condition for the effective functioning of such a tourist centre. The system comprises tourist infor-mation offices, tourist offices and also information points operating in hotels. Apart from their usual range of duties, hotel staff provide tourist information and sell tickets. Such information is available in each hotel. The tourist offices ensure transmission of tourists from Quebec to the tourist centre. It is important that the tourist centre should function not only in the summer period but all year round.

Describing all the attractions of the Côte-de-Beaupré centre would take too much space. Therefore, only a few information items are given – about the Mont-morency Falls, Mary's Farm, the Copper House and St Anne Basilica.

We shall start the trip around Côte-de-Beaupré from the Montmorency Falls, which is on the borderland of Quebec. It is a big waterfall, one of the biggest in Canada. Watching the waterfall is free of charge. It can be accessed for free because it is treated as national property. However, around the waterfall there have been built accompanying facilities that present such encouraging offers that the tourist feels in a way impelled to purchase some products or services there. The first such attraction is the cable car, whose route runs near the waterfall. For a small charge one can admire from a short distance the masses of water falling down almost right into the St Lawrence River. At the upper and lower terrace there are souvenir shops, a restaurant and a hotel. There are also tourist paths that enable tourists to get right under the waterfall.

Another attraction of Côte-de-Beaupré is Albert Gilles's Copper House. The Copper House was created by several generations as a family business. Artistic copperplates are made and sold there. The family members have dealt with making copperplates for several generations. Currently the facility is a workshop, shop, museum and exhibition at the same time. A large gallery of beautiful copperplates first encourages the tourist to look at them and then tempts them to buy them. It is also possible to see a reconstruction of a gallery from which copper is mined. Copperplate making is not an agritourist element, but the house is perfectly matched with the tourist centre system, which makes the centre even more attractive so that tourists are encouraged to visit it.

Mary's Farm is the third attraction. It is the dwelling house of an old farm, where one function has been left – home bread baking, which the owner and her family do. The bread is sliced and with maple syrup poured over it is sold to tourists. The tasting of the bread is an expensive pleasure, but there are a number of ready consumers. Admission to the farm and watching the bread baking is free of charge. In addition, in the dwelling house a shop has been opened where maple syrup and many local products manufactured by the local farmers are offered.

Another attraction is the large national pilgrimage centre. It consists of several sacred buildings, including St Anne's Church. The church houses the miraculous statue of St Anne, the mother of Saint Mary. Many years ago, healings started at the local church. As a result, pilgrimage traffic on a large scale developed. A large cathedral was built in the place of the former small church. The pilgrimage centre maintains a close link with the whole tourist centre and ensures a regular inflow of clients to Côte-de-Beaupré all year round.

The United Kingdom

Mature agritourism

The United Kingdom has a long and rich experience in the development of tourist enterprises in rural areas. Its agritourist offer is extremely diversified and it is still being enriched with more and more new products. The whole of the United Kingdom has excellent conditions for agritourism, perhaps with the exception of the central and south-eastern part of England, which is lowland. Despite its predisposition and beautiful scenery in Northern Ireland tourism is least developed because of the long-lasting conflicts between the aboriginal Irish and the Protestant Scots. When comparing Northern Ireland with the Irish Republic, it is clear how political conflicts and fighting make the development of tourism difficult. While in Ireland tourism has been flourishing, this is not the case in Northern Ireland. Almost every region in the rest of the United Kingdom can boast great examples of the development of agritourism. The south-western part of the United Kingdom, called the West Country, the centre – Yorkshine Dales and Northumberland – and Scotland have excellent infrastructure for tourism in rural areas. Many examples of agritourist enterprises described in this book come from the United Kingdom.

Geographical context

The United Kingdom is situated in Great Britain, in the north-eastern part of Ireland and on many islands on the continental shelf of western Europe and it comprises England, Scotland, Wales and Northern Ireland. There are 59.07 million British citizens (2002) inhabiting an area of 242.9 thousand km^2, with 94.2% Britons and 1.4% of Indian extraction. Also, there are Pakistanis, Bangladeshis, Chinese, Africans and others inhabiting the country. Scotland and the islands situated near its coast, the Hebrides, Orkneys and Shetlands, are the northernmost areas of the United Kingdom (Bateman and Egan, 2000). In Scotland three major regions can be distinguished: the Caledonian Mountains and Grampians in the north, the lowlands in the centre and the Southern Uplands in the south. The highest peak of Scotland is Ben Nevis (1344 m) in the Grampians.

The Cheviot Hills form a natural boundary between Scotland and England. Further south there is the range of the Pennines running parallel to the coasts across the Midlands. In the north-western part of England there are the Cambrian Mountains and the Lake District. In the county of North Yorkshire in north-eastern England there are peat bogs with heather growing on them.

South of the Pennines stretch the undulating hills of the Midlands, contrasting with the plain areas of the eastern part of the country. The longest river in the United Kingdom – the Thames – rises in the Cotswold Hills, runs through the centre of London and flows into the North Sea. Western England borders on Wales, which is surrounded by the waters of the Irish Sea in the north, St George's Channel in the west and the Bristol Channel in the south. The Cambrian Mountains rise in the central part of Wales. Ulster is the part of Ireland that belongs to the United Kingdom. North of Belfast – the capital of the province – there are the Antrim Mountains, whose steep slopes of chalk and basalt fall into the water and produce the south-western coast of Scotland.

Poland

A poem on Polish agricultural space

The agricultural and rural space of Poland has been extremely attractive for centuries. A 19th-century national lyricist, Adam Mickiewicz, a major Polish poet, describes the space in a particularly beautiful manner in the national epic *Pan Tadeusz* (Mickiewicz, 1990). Below are two quotations. The first is a description of a home garden and the other describes a landscape of fields at harvest.

> It was the kitchen garden. Row on row
> Of fruit-trees give their shade to beds below.
> A cabbage sits and bows her scrawny pate,
> Musing upon her vegetable fate;
> The slender bean entwines the carrot's tresses
> And with a thousand eyes his love expresses;
> Here hangs the golden tassel of the maize,

And there a bellied water-melon strays,
That rolling from his stem far off is found,
A stranger on the crimson beetroots' ground.

(Book II: The Castle)

The reaping women had begun their song,
As mournful as a rainy day and long,
That in the dampening mist more sadly died;
The sickles swished and all the meadow sighed.
Behind a line of mowers moved along
Cutting the aftermath with endless song,
A whistling song, and at each verse's close
They stopped to whet their scythes with rhythmic blows.
The mist hides all – and all the sounds conspire,
Song, scythe and sickle in a hidden choir.

(Book VI: The Village of the Gentry)

Regional diversification of agritourist environment

Poland is located in Central Europe. In 2007 the total area was 312 thousand square kilometres and the population was 38.1 million. It is estimated that at least 70% of the country is suitable for the development of agritourism. Tourist areas are generally uniformly distributed throughout the country. However, there are regions in Poland, such as Podlasie and Masuria, Pomerania, the Lubuski Land, northern parts of the Wielkopolska region, the Carpathians, the Sudetes and Roztocze, that exhibit unique agritourist potential. A characteristic feature of Poland is its parallel belted agricultural space. The northern belt is the Baltic Sea coast and lake belts with plenty of forests and lakes (see Figs 18.4 and 18.5). The central plateau belt is covered predominantly by agricultural land. The south and south-west are covered by hills, with the Tatra mountains in the very south (Fig. 18.6). In the middle of the country, from south to north flows the River Vistula dividing Poland into eastern and western parts. The landscape is diversified in respect of area configuration, type of development, size of farms and type and specialization of production. Considerable differences occur between individual regions of the country. The areas that are more conducive to the development of agritourism than others are above all those with altitude differences, hilly or mountainous terrains, places with rivers or lakes. Places with interesting architecture also diversify agritourist space. There are many unique, beautiful places.

Kazimierz Dolny on the River Vistula is one of many attractive places and an instance of terrain that is conducive to the development of agritourism. The loess ravines are unique as they cannot be found anywhere else in Poland. The Vistula Canyon, with considerable altitude differences, and the Kazimierz architecture make this area more suitable for agritourist activity than others. Additionally, the several-century-old traditions of the local people, both Polish and Jewish, are also used for the development of tourism. The use of folk baking products and selling them as tourist attractions deserve special attention (Fig. 18.7).

There are several traditional, historically recognized regions with different agritourist spaces and environments and considerable variation. The diversification

Fig. 18.4. Diversification of agricultural space of Poland from north to south: Baltic sea coast and sandy beach (photo by M. Sznajder).

Fig. 18.5. Diversification of agricultural space of Poland from north to south: the Wielkopolska region – typical landscape, lake, forest and fields of crops (photo by M. Sznajder).

Fig. 18.6. Diversification of agricultural space of Poland from north to south: Tatra mountains (photo by M. Sznajder).

Fig. 18.7. A cockerel and other regional baking products of Kazimierz Dolny are popular tourist purchases (photo by L. Przezbórska).

of the agritourist environment of Poland is indicated with a brief description of three different provinces: Greater Poland, Lower Silesia and Podlasie.

Wielkopolska (Greater Poland)

The agritourist environment of Wielkopolska has a unique character that distinguishes it from other regions of the country (Fig. 18.8). It is not so diversified in respect of area configuration as the southern and south-western regions of Poland. The terrain is usually flat; only in the north-western part of the province is it more undulating. The most attractive physiographic forms of Wielkopolska are the valleys of the rivers and the lake districts.

In Wielkopolska there are relatively few endemic and rarer species of fauna and flora with the exception of an insect stag beetle (*Lucanus cervus*) in Rogalin and resettled animal species such as beavers, wolves and European bison. The landscape is dominated by crops on arable land, and the share of permanent grassland, with the exception of river valleys, is small. In the northern part of the province the forest space and lakes play an important role. Most towns were located in Magdeburg Law however, the south-eastern part is characterized by dispersed development. In the south of Wielkopolska small and medium farms are dominant. Moving to the north of the province the farms are bigger and bigger. In the central part of Wielkopolska big farms can be found. Their share in the north of the province is considerable. Despite the considerable share of animal

Fig. 18.8. Rural community heritage of the Wielkopolska region: wooden churches are a very common element of the architecture of rural areas (photo by M. Sznajder).

production, the agritourist space has an agricultural character. In spring, summer and autumn grazing animals cannot be seen, which is characteristic of many parts of Poland, as the cattle are usually kept indoors. The landscape is dominated by cereals, potato, sugarbeet and rape plantations. In certain parts of the province there are enclaves of a horticultural or orchard type. Hothouses dominate the landscape north of Kalisz. Wielkopolska has a significant capacity for the development of agritourist farms, especially on lakes and in forest areas. An important contribution to their development is the construction of a motorway, which will enable the inhabitants of Berlin easier access and will probably ensure a considerable and steady inflow of tourists to the area.

Lower Silesia

This has an attractively configured space due to the biophysical characteristics. More rare species of animals and plants can be found here. However, the utilization of land seems to be more monotonous than in Wielkopolska. Owing to the disappearance of dairy farming, the cropping activity excludes fodder plants. Lower Silesia is characterized by distinct, quite desolate rural buildings. The differences are noticeable almost immediately on entering Lower Silesia from Wielkopolska. The area of Lower Silesia changes from undulating in the northeast and east through plains in the centre to mountainous in the south-west.

In contrast to Wielkopolska, Lower Silesia lacks natural lakes almost completely. However, the valleys of the River Oder and minor mountain rivers are picturesque. The Kłodzko Basin, surrounded by mountains on each side, is especially attractive. The use of land in Lower Silesia correlates with the terrain configuration. In the lowlands there are mainly arable lands and in the mountains forests. Owing to the lack of sheep and cattle, mountain pastures are increasingly converted into forest areas. Near Milicz there are vast forest expanses and fish ponds. Farm buildings are usually dilapidated. In many villages buildings have not been renovated since the end of the Second World War. The farmers, who were mainly obligatorily displaced persons from the eastern borderland, did not renovate them because they did not consider themselves to be their owners. The farms in Lower Silesia are mainly orientated towards plant production. Figure 18.9 shows a diagram of a typical plan of a farm in the Lower Silesia province. The development of farms tends to correspond to the farming conditions at the beginning of the twentieth century.

Lower Silesia is attractive to agritourism. This particularly concerns the whole range of the Sudeten Mountains, but also the Trzebnica Hills. The motorway that runs across the centre of Lower Silesia makes for an easy inflow of tourists from western Europe.

Podlasie

This province lies in north-eastern Poland. In the south the terrain is rather flat, but it gradually becomes more undulating as we move north, especially near

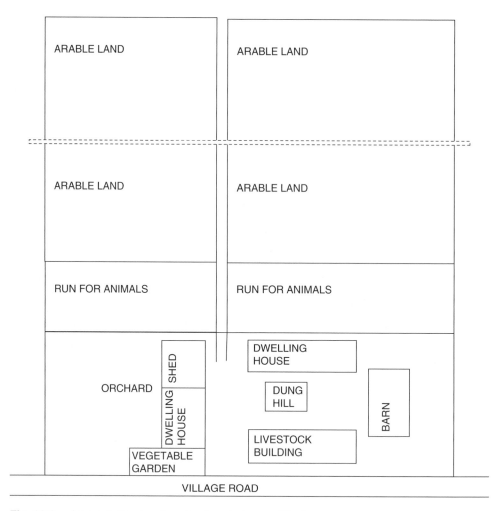

Fig. 18.9. An example of a plan of a farm in Lower Silesia.

Suwałki. The province has numerous interesting and scenic locations, such as the Bug Canyon or the Biebrza Valley. There are also plentiful lakes near Augustów and Suwałki. In the south of the province arable lands dominate. As we move north, the area of meadows and pastures increases, then that of forests. In the northernmost part arable lands dominate again.

The ethnographic variety of the province is worth noting (Figs. 18.10 and 18.11). A large part of the population is Orthodox, but there are also a few Old Believers, an unreformed faction of the Orthodox Church. The area of the province is also inhabited by the Tartars, who are Muslims. Alongside the Poles there are also Belarussians and Lithuanians. The development is diversified, though the dispersed type prevails. There are unique plant communities and animals such as European bison, elk, lynxes, sandpipers, black herons, etc. For this reason as many as four national parks have been established as well as many scenic

Fig. 18.10. Rural community heritage of Podlasie province: Orthodox abbey (Bazillians) on a Supraśl river bank (photo by M. Sznajder).

Fig. 18.11. Rural community heritage of Podlasie province: the Jewish culture – a synagogue in Tykocin (photo by M. Sznajder).

parks. The province is especially predisposed to the development of agritourism, which has a seasonal character. The distance from big cities and the absence of a motorway are the main constraint to the tourist development of the region. In order to make the region more widely available to tourists, huge investments need to be made.

Case Studies

Try to present your country, region or province in a promotional flyer or website as a unique place of agritourist attraction. Complete the task either for a holiday-maker expecting relaxation or for a demanding agritourist. Think how attractive your description could be to the reader.

IV Conclusions

19 Agritourism Yesterday, Today and Tomorrow

Is Agritourism Still Agritourism?

At the end of the book a question returns: *What actually is agritourism?* The term agritourism has settled in the colloquial language of many nations. However, because it is relatively new, the range of meaning is not yet stable.

An average city inhabitant associates agritourism with a farm offering accommodation and not clearly specified relaxation. With time and with the development of agritourism itself, the colloquial meaning of this term has become extended. This results from the fact that more and more people have personal experience of agritourism and they pass it on to others. It is the extension of the range of meanings associated with the term agritourism and the development of appropriate emotional images in colloquial language that have enormous importance for the further development of agritourism. Its positive or negative connections result from specific personal tourists' experience and from the manner of promotion of agritourism.

In specialized literature the term agritourism is more stable than in colloquial language. However, even here the process of its maturing has not finished. There is a difference between the semantics of agritourism seen from the point of view of an agritourist and from the point of view of a provider of agritourist services. For the tourist, agritourism means all forms of tourism and relaxation related to agriculture, animal breeding and food production. In this case the natural process is an extension of the meaning by ethnography, which includes the cultural, social and religious activity of farmers and rural communities. Thus, increasingly, other forms of tourism in rural areas are included in the range of the term agritourism. For the farmer, agritourism means different forms of supplementary or exclusive business activity on a farm, primarily in order to increase income.

The extension of the meaning of agritourism is an objective process. Long ago it crossed the limits of the original range given to the term. At the moment it also encompasses psychological, ethnographic, economic and social aspects. A fundamental question arises: *Is agritourism still agritourism?* Perhaps it is an exclusive term into which everything happening in tourism related to agriculture, rural areas and nature is included. Or maybe it is a comfortable substitute term for a dynamically developing new branch of business that has not been described yet and does not yet have an appropriate name.

Immersion in nature is man's natural and internal need

People have numerous needs, which they try to satisfy. One of them is the need of contact or even immersion in nature. This need is a basic human need. The need of immersion in nature did not find a proper place in Maslow's classic theory of needs (Maslow, 1943). It is obvious, natural and internal and thus it is often unconscious. For centuries people have satisfied this need by working in agriculture, horticulture, animal husbandry or forestry and even in hunting. Because the people working in these professions are in a way naturally immersed in nature, they often do not realize the fact that contact with it is a human need. City inhabitants find it easier to notice that human contact with nature is necessary. Therefore, zoos, parks, playgrounds, botanical gardens, palm houses and orangeries were established in cities, as they bring people closer to nature.

City inhabitants individually satisfy this need in a variety of ways. For example, it is common to keep dogs, cats, guinea pigs, canaries and other animals in city houses and apartments. Many people have small gardens near their houses or grow ornamental plants on balconies or terraces and in homes. Tourist traffic and holidaymaking away from urban areas are developing. These attempts are not sufficient, because for full development people need to have an even closer contact with nature, farming and countryside.

Urbanization as a reason for the loss of contact with nature

The growth of the urban population is a common tendency around the world. According to forecasts, by 2030 the urban population will have grown to nearly 5 billion people and will be about 60% of the total population. The percentage of rural population is constantly decreasing. At the moment there is a balance between the rural and urban population, but soon the urban population will definitely outnumber the rural population. According to demographic forecasts, by 2030 the rural population will have decreased to about 3.3 billion. There will be even fewer farmers, because many village inhabitants will stop being active in farming. Urbanization means mass loss of contact with nature and farming around the world. The loss will be seen in limited time spent outdoors in the fresh air and almost complete loss of contact with farming, animal breeding and food production processes.

City inhabitants have definitely reduced the time spent outdoors in comparison with rural inhabitants. Their work is done mainly indoors and after work they also spend time indoors – in flats, shops, restaurants, pubs, cinemas, etc. It is possible to make interesting observations on a warm, sunny, spring day in every city living quarter, where thousands of people live. On such a nice day many people would benefit from being out in the fresh air, but they are not. This means that they spend time inside. A city inhabitant needs to make a certain kind of trip to reach a park whereas a farm inhabitant has nature on hand. In practice, contact with agriculture, forestry and rural areas has been completely lost. Many city dwellers do not have the faintest idea how milk, cheese, bread and sugar are made or where they come from. City inhabitants' attitude to natural smells has changed. Often smells that used to be regarded as nice are now considered unpleasant and repulsive. The loss of touch with nature causes various negative consequences for an individual city inhabitant and whole urban communities. This applies to problems of physical and mental health, emotions, social pathologies and the knowledge of nature and especially food. Agritourism is a method of counteracting these negative results.

Objective and permanent reasons for the development of agritourism

The development of agritourism is already widely recognized. However, a question arises: *What has been the most important reason for its beginning and development?* So far, above all it has been noted that the most important reason for the development is the need to increase farmers' income. Government programmes, being based on this reason, have supported farms involved in the development of agritourism. Agritourism as an additional agricultural activity is supported especially by the governments of the USA, Canada and the European Union. In many wealthy countries there is particularly strong government support. In poorer countries, which are distant from economic centres, this is still a hidden potential, which will develop rapidly in favourable conditions. Without negating the importance and sense of this reason, it is possible to put forward a new hypothesis: that the main driving force behind the development of agritourism is the necessity to satisfy people's natural and internal need to have contact with nature, which has been lost in consequence of urbanization. It is the need of human contact with nature that is the ultimate cause of the origin and development of agritourism. This is a permanent need, so agritourism is potentially sustainable. The hypothesis that is put forward shifts the argument from the need to develop agritourism from the supply side, i.e. farmers, to the demand side, i.e. city inhabitants and non-farmers. In this way it both considerably broadens the base of those interested in the development of agritourism (city inhabitants) and reinforces the social argument for supporting it.

Substantial potential demand for agritourism

If we classify agritourism as a basic human need to be satisfied and if we take rapid urbanization into consideration, we should expect an enormous demand for agritourism in consequence of the effect of these factors. In practice, the potential demand for agritourism seems to be large. A question arises: *Is world agriculture ready to satisfy the demand?* For the time being, the supply of agritourist services is still larger than the demand, but the demand is constantly growing. When considering the interaction of supply and demand it is necessary:

- First of all, to realize that this is a potential demand of which many people are still unaware. If it were fully activated, the market would be enormously unbalanced. The activation of the potential will be gradual, but its growth tempo will be rapid.
- Second, a relatively limited tempo of the growth of income of urban population will hinder the rapid growth of demand for these services.
- Third, the market mechanism balances demand and supply. This means that prices of agritourist services will be constantly growing along with the rising number of agritourists. From relatively cheap services they will be transformed into more and more expensive services, which only a limited number of tourists will be able to afford.
- Fourth, new mass and valuable agritourist products and services will also be developed, which will satisfy the demand of less wealthy consumers or those who wish to spend less money on them.

Ample offer of agritourism

When studying literature on the subject and when visiting specific agritourist farms one can see the wide variety of products and services offered. This book presents their classification, which confirms the variety. Nine categories of products and services have been distinguished and further, more detailed subcategories within each category. They could be even further diversified. The literature provides a surprising diversity of agritourist products and services. New concepts are constantly being developed, which extend the variety. There is an ample offer of agritourist products and services. Without taking a risk, it is possible to put forward a hypothesis that the number of products and services that agritourism can offer is unlimited. Such diversity results in the uniqueness of each agritourist farm. The driving force behind the attractiveness of agritourism is its unlimited possibility to create new and increasingly surprising products. The next decades will abound in new, surprising and attractive solutions. Some of these products give a basis for distinguishing product agritourism that is related to a particular product, e.g. wine tourism is related to wine and dairy tourism is related to the dairy industry.

Agritourism closer to cities

Modern tendencies of development indicate that agritourism cannot be exclusively associated with farms. There are agritourist products and services that can be offered outside a classic farm. Instead of serving individual clients, there are agritourist services for mass consumers. Maize labyrinths or horticultural centres are good examples of this. Probably soon there will be newer and more advanced agritourist products for mass consumers. In the past, agritourism involved a trip to the country or a farm. Today the situation is changing. The marketing rule that says that the product should be available where the client wants it is slowly coming into practice. Instead of transporting mass tourists to distant rural areas, services will be provided for them on city outskirts or even in centres. More and more frequently, entities providing agritourist services settle there so that clients can have quick and easy access to them – for example, the development of agritourist farms aimed at cooperation with schools and nursery schools within the 'green school' programme, which are located around big cities.

Widespread geographical availability of agritourism

Agritourism can actually be offered and practised in every corner of the world, because there are farms everywhere. After all, food production and processing are basic human activities. The widespread character of agritourism is proved, for example, by the case studies given in this book, which come from 20 different countries of the world. Agritourist business is also made in other countries, which is evident in numerous publications or advertisements on the Internet, for example. Naturally, some areas are particularly predisposed to the development of agritourism. This applies to areas situated in picturesque surroundings, with unpolluted air and water. However, other areas that might seem to be less attractive may also be places of interesting agritourist services. For many farmers thinking of agritourism, it is a serious problem to realize the uniqueness of the location and attractiveness of their farms. They ask themselves a question: *What could be interesting about my ordinary farm?* However, what could be ordinary and unattractive for a local may be unique for an agritourist. Let us imagine the attitude to a rice paddy of a South-east Asian farmer and a European tourist. For the former it is nothing interesting, whereas for the latter it is a great attraction to see. Recognizing the uniqueness of one's own farm and nearest surroundings is one of the major challenges in the development of agritourism. The feeling that a farm is of banal character, thinking that there is nothing interesting in it for an agritourist, is a serious hindrance to the development of agritourism.

Agritourism diversifies the world

The 20th century is considered to have been the age of specialization, concentration, increased efficiency of work and unification of products and services. These tendencies result from the theory of production scale. Agritourism is a phenomenon initiating new trends in modern economics, which will soon grow stronger. Agritourism results from diversification and it is oriented towards diversification of the world. This attitude can be justified by several examples:

- Agritourism is becoming a channel of direct sales of local products, made in small amounts. In this sense it diversifies what is unified by supermarkets and large processing companies. Agritourist farms produce and then sell tourists short batches of specific products, which will not be found on supermarket shelves and which will not be an object of interest to the processing industry.

- In the sense of production, the agritourist farm wants to achieve a diversity of character both in plant and animal production (many species of plants and animals) and in processing. Commodity farms, cooperating with the powerful processing industry, aim at specialization and increased scale of production.
- Agritourism diversifies and personalizes agritourist services, whereas in mass tourism there is increasing unification and standardization.

These examples of the diversifying role of agritourism do not eliminate the danger of the opposite processes, i.e. unification and increasing scale. Big capital, which is looking for investment possibilities to maximize profits, has taken control of some agritourist businesses, where the production scale and effectiveness are indexes of primary importance.

Agritourism softens the effects of increased work efficiency

The second half of the 20th century was also a period of the clashing of two opposed processes in agriculture. The first process, also in a historical sense, concerns the rapid increase in work efficiency in agriculture. At the beginning of the 20th century one person employed in farming produced food for two to three people, whereas at the moment in well-developed countries one person produces food for as many as 70–100 people. Increasing work productivity caused a considerable decrease in the demand for labour in agriculture. Rural inhabitants began to face the dilemma of how to find means to support themselves and their families. For some of them the solution was to find a job in the city. For this reason rural areas are becoming depopulated. In order to keep people in rural areas state governments more and more frequently promote non-agricultural development of the areas and, for example in Canada (Manitoba and Saskatchewan), they promote agriventure. Agritourism perfectly fits the concepts of the multifunctional development of rural areas.

The other process, which is delayed in a historical sense in comparison with the first and whose effects are opposite to those of the first, is increasing tourist traffic in rural areas. In order to handle this traffic it is necessary for people to stay in rural areas and in some cases to return to the country.

The two processes are asymmetrical. This means that the reduction of workplaces in agriculture is still larger than the number of workplaces offered by agritourism. However, soon it may turn out that the importance of the latter process will grow. This results from the constant increase in urban population around the world. Urban people are increasingly interested in visiting rural areas and the farms situated there. They are also increasingly interested in information about how food is made. They want to stay close to plants and animals. This need is arguably an immanent human attribute. Simultaneously, people around the world are getting richer. Therefore, they will be ready to spend a considerable amount of their income on tourism, including agritourism.

From the Depths of Centuries

When and where did agritourism begin? This question takes us into the depths of centuries without a particular date. This should rather be linked to the history of development of urban societies and some people abandoning farming, pasturing, hunting and gathering. Arguably the oldest city in the world is Jericho, which was founded about 9000 BC, so 'agritourism' might have an extremely long history. Naturally, this is not agritourism in the current meaning of the word. This was a completely different phenomenon. In the past centuries a stranger's visit to rural areas was a great event and blessing. He could count on local people's favour and hospitality. He brought news from the rich and famous world, which all villagers waited for. However, he may also have been treated as an intelligence agent and spy. For centuries people living in cities have contacted their relatives living in the country. They have sent their children there and agreed on compensation for their stay. This pattern has concerned many regions of the world. On the other hand, for example, in ancient times the Romans travelled to rural areas to relax in their holiday homes.

The 19th century brought enormous changes in 'agritourism'. It was then that city dwellers began to follow the fashion to go on holiday in the country. It was the development of the railway, which started in 1825, that made trips to the country much easier. Historians place the first national initiatives of agritourism in the period following the beginning of rail transport.

Since the 1920s the development of the motor industry has given a new, better chance for city inhabitants to visit the country. Rural inhabitants were more and more open to the possibility of extra income resulting from the handling of tourism. In the 1970s and 1980s different new forms of services provided for tourists in rural areas began to develop. The 1990s and the beginning of the 21st century were a period of recognizing an increasing value of agritourism. At the moment we are facing a real agritourist boom.

Agri-cybertourism on Hand

Finishing the book about agritourism, we shall try to look at agritourism in the future. From historical considerations we can conclude that the development of agritourism is mainly a function of the development of mass transport and information. It is these factors that make contact with nature possible, which is man's natural need. Constructing a pattern for the agritourism of the future does not require unrealistic fantasizing.

When constructing this model, the first question arises: Should we expect such an important revolution in transport as the one related to rail, car and air transport? It does not seem that man will invent a new means of public transport in the near future. In fact, we are left with the present means of transport, but much faster and more comfortable and with a bigger capacity. They will enable easier, faster and cheaper transport of a larger number of tourists to distant corners of the world. In order for this to happen, it is necessary to develop new fast railways, motorways, airports and airlines. Increasing travelling time and comfort will cause a statistical tourist to use agritourism many times a year. However, not everybody will be able to go to the country, for example, those on low incomes. Forms of agritourism in the city will be developed for them.

The second question concerns the influence of new information technologies on agritourism. We should expect great changes in this field. They will be made by the Internet, the GPS, e-banking and methods of global reservations. A future agritourist will use a 'Global Agritourist Planner' – named AGRICYBER-TOUR, for example. On the map the planner will circle the area the tourist would like to visit; the search engine will give the parameters of farms to stay in and a holiday programme. It will ask about the planned time of visit and preferences concerning accommodation and board, and AGRICYBERTOUR will do the rest of the tasks automatically. It will find an appropriate farm, make a plan of the visit and reservations, store the necessary numbers in our telephone, make all payments and buy insurance from our account, and in an emergency it will alarm the right people and services. AGRICYBERTOUR will provide the farmer with full information about the person planning a visit, the arrival and departure time and services to use and, most importantly, it will guarantee income for the services provided. Agri-cybertourism is already on hand.

References

Altkorn, J. (2000) *Marketing w turystyce* (Marketing in Tourism). PWN, Warsaw.

Barthelemy, P.A. (2000) *Changes in Agricultural Employment.* The European Commission, Agriculture and Environment, Brussels, Belgium. Available online at http://europa.eu.int/comm/agriculture/index.htm, accessed April 2008.

Bateman, G. and Egan, V. (2000) *Geografia państw świata* (Geography of the Countries of the World). Wydawnictwo MUZA, Warsaw.

Becken, S. (2003) Reducing the dependency on tourist icons and why it is important for sustainability. In: *Proceedings of the Conference Taking Tourism to the Limits,* 8–11 December 2003. Department of Tourism Management, University of Waikato Management School.

Begg, D., Fischer, S., Dornbusch, R.v (1995) Ekonomia (Economy), volume 1, Polskie Wydawnictwo Ekonomiczne, Warsaw.

Blevins, J. (2003) More Colorado farmers dabble in 'agritainment'. *Knight Ridder Tribune Business News,* Washington, 24 November.

Bruch, M.L. and Holland, R. (2004) *A Snapshot of Tennessee Agritourism: Results from the 2003 Enterprise Inventory.* University of Tennessee, Centre for Profitable Agriculture. Available online at http://cpa.utk.edu/pdffiles/PB1747.pdf, accessed April 2008.

Bryden, J.M. (2000) Structural changes in rural Europe, In: *Western Agri-Food Institute Colloquium on Rural Adaptation to Structural Change: Summary Paper and Critique, May–June 2000, An On-line Colloquium on Structural Changes in Rural Areas of America, Australia, Canada, and Europe.* ftp://132.205.182.30/western-agrifood_institute/rural_restructuring_europe.pdf

Butler, R.W., Hall, C.M. and Jenkins, J. (1998) *Agritourism and Recreation in Rural Areas.* John Wiley and Sons.

Cohen, E. (1972) Toward a sociology of international tourism, *Social Research,* 39, 164–82.

Cooper, C., Fletcher, J., Gilbert, D., Shepherd, R. and Wanhill, S. (2005) *Tourism: Principles and Practice.* Longman, Harlow, UK.

Daneshkhu, S. (2001) Foot-and-mouth hits tourism. *Financial Times,* 11 May.

Davis, J.H. and Goldberg, R.A.A. (1957) *A Concept of Agribusiness.* Alpine Press, Harvard University, Boston, 136 pp.

Denman, R. and Denman, J. (1993) *The Farm Tourism Market*. The Tourism Company, Ledbury, UK.

Dinell, D. (2003) *Wichita Business Journal,* 9 May, 18 (19), 3.

Drzewiecki, M. (1992) *Wiejska przestrzeń rekreacyjna* (Rural Recreational Space). Instytut Turystyki (Institute of Tourism), Warsaw.

Eckert, J. (2004) *Growing Agritourism. A Starter Manual for Farmers and Ranchers.* Eckert Agrimarketing, St Louis, Missouri, USA.

Embacher, H. (2003) Quality classification of farms in Austria. Presentation at the workshop 'Quality in the European Rural Tourism' organized by the Leonardo-da-Vinci Project, Puchheim, 16 May.

Fisher, A.G.B. (1935) *The Clash of Progress and Security*. Macmillan, London.

Fourastié, J. (1949) *Le Grand Espoir du XXe siècle* (The Great Hope of the Twentieth Century). *Progrès technique, progrès économique, progrès social.* Presses Universitaires de France, Paris, 224 pp.

Frost, W. (2003) Bravehearted Ned Kelly. In: *Proceedings of the Conference Taking Tourism to the Limits,* 8–11 December 2003. Department of Tourism Management, University of Waikato Management School, Hamilton.

Gaworecki, W.W. (2006) *Turystyka* (Tourism). Polskie Wydawnictwo Ekonomiczne (Polish Economics Publishing House), Warsaw.

Geisler, M. (2008) *Agritourism Profile*. AgMRC, Agricultural Marketing Resource Center, Hawaii Agricultural Statistics Service. Available online at http://www.agmrc.org/agmrc/commodity/agritourism/agritourism/agritourismprofile.htm, accessed April 2008.

Getz, D. and Page, S. (1997) *The business of rural tourism: international perspectives.* Thomson Learning EMEA, Auckland. *New Zealand Herald*, 26 November 2003.

Halamska, M. (2001) Obecne i przyszłe zróżnicowanie regionalne wsi (Present and future regional differentiation of the countryside). In: Kolarska-Bobińska, L., Rosner, A. and Wilkin, J. (eds) *Przyszłość wsi polskiej. Wizje, strategie, koncepcje* (The Future of the Polish Countryside: Visions, Strategies and Concepts). Institute for Public Affairs, Warsaw, pp. 51–62.

Hall D., Mitchell M. and Roberts L. (2004) *New Direction in Rural Tourism*. Backer and Tylor Books, Ashgate Publishing Ltd., Aldershot, UK.

Hall, D.R., Kirkpatrick, I. and Morag, M. (eds) (2005) *Rural Tourism and Sustainable Business (Aspects of Tourism)*. Multilingual Matters, Clevedon, UK.

Hall, M.C., Sharples, L., Cambourne, B. and Macionis, N. (2000) *Wine Tourism around the World*. Butterworth Heinemann.

Hall, M.C., Sharples, L., Mitchell, R., Macionis, N. and Cambourne, B. (2003) *Food Tourism around the World*. Butterworth Heinemann.

Iakovidou, O., Partalidou, M. and Manos, B. (2000) Rural tourism. Agritourism: a challenge for the development of the Greek countryside. In: *International Seminar: Agritourism and Rural Tourism. A Key Option for the Rural Integrated and Sustainable Development Strategy*, pp. 65–70, September 21–22, 2000. Informal International Association of Experts in Rural Tourism (I.I.A.E.R.T.), University of Perugia, Agricultural Faculty, Center for Agriculture and Rural Development (Ce. S.A.R.), Scientific and Cultural Association Biosphera, Perugia, Italy.

Jenkins, J., Hall, C.M. and Troughton, M. (1998) The restructuring of rural economies: rural tourism and recreation as a government response. In: Butler, R., Hall, C.M. and Jenkins, J. (eds) *Tourism and Recreation in Rural Areas*. John Wiley and Sons, Chichester, UK.

Jolly, D.A. and Reynolds, K.A. (2005) Consumer demand for agriculture and on-farm nature tourism, US Small Farm Center Research Brief, University of California, Davis. Available online at http://www.sfc.ucdavis.edu/agritourism/agritour.html, accessed April 2008.

Kamiński, W. (1995) Warianty wielofunkcyjnego rozwoju wsi – uwarunkowania przestrzenne (Variants of multifunctional rural development – spatial conditions). In: *Wielofunkcyjny rozwój wsi w aspekcie przezwyciężania przeludnienia agrarnego* (Multifunctional Rural Development in the Context of Overcoming Agrarian Overpopulation). Scientific Papers of the Agricultural University of Cracow, No. 295, issue 43. Agricultural University Publication House, Cracow, pp. 19–25.

Kay, R.D., Edwards, W.M. and Duffy, P.A. (2007) *Farm Management.* McGraw-Hill Science/Engineering/Math, New York.

Keeney, M. and Matthews, A. (2000) Multiple job holding: explaining participation in off-farm labour markets. Labour demand and labour supply of Irish farm households. Poster presented at XXIV International Conference of Agricultural Economists, Berlin.

Kotler, P. (1999) *Marketing. Analiza, planowanie, wdrażanie i kontrolowanie* (Marketing, Analysis, Planning, Implementation and Controlling). Felberg SJA, Warsaw.

Krapf, M. (2003) *Zanzibar – goździkowa wyspa* (Zanzibar – the Clove Island). Kurier PKP No. 93, Kolejowa Oficyna Wydawnicza, Warsaw.

Lane, B. (1992) A review for the Organization for Economic Co-operation Development. Mimeo, Rural Tourism Unit, Department for Continuing Education, University of Bristol.

Łysko, A. (2002) Zmiany w krajobrazie rolniczym na obszarze moreny czołowej w okolicach Czaplinka w latach 1948–1998 (Changes in rural landscape in the area of the end moraine near Czaplinek in the years 1948–1998). Doctoral dissertation, Akademia Rolnicza (Agricultural University), Szczecin.

Mahé, L.P. and Ortalo-Magné, F. (1999) Five proposals for a European model of the countryside, CAP and the countryside, proposals for the food production and the rural development. In: *Economic Policy, a European Forum.* No. 28/1999, Centre for Economic Policy Research, Centre for Economic Studies, Maison des Sciences de l'Homme, Blackwell Publishers, London, UK.

Majewski, J. (2001) Produkt. Identyfikacja produktu – maszynopis (A product. Identification of the product. Manuscript). Akademia Rolnicza, Poznań.

Majkowski, K. (2004) Klasyfikacja i częstość produktów i usług agroturystycznych na podstawie oferty internetowej gospodarstw Polski, Włoch, Wielkiej Brytanii i Stanów Zjednoczonych (The classification and frequency of agritourist products and services in the example of the Internet offer of farms in Poland, Italy, the United Kingdom and the United States of America). Masters thesis, Agricultural University of Poznań.

Maslow, A.H. (1943) A theory of human motivation. Originally published in *Psychological Review* 50, 370–396.

Mayse, J. (2003) Lamily to launch Ovensboro, Ky. – Area Educational Corn Maze. *Knight Ridder Tribune Business News, Washington*, 22 March.

Mickiewicz, A. (1990) *Pan Tadeusz* (Mr Thaddeus), trans. MacKenzie, K.R. Polish Cultural Foundation, London.

Middleton, V.T.C. (1996) *Marketing w turystyce* (Marketing in Tourism). Polska Agencja Promocji Turystyki (Polish Agency for the Promotion of Tourism), Warsaw.

Raciborski, J., Turlejska, H., Zielińska, R., Kmita-Dziasek, E., Śliz, J., Rzutki, M. and Jasiński, J. (2005) *Prawno-finansowe uwarunkowania prowadzenia usług turystycznych na polskiej wsi po akcesji do UE.* Poradnik Praktyczny, Centrum Doradztwa Rolniczego w Brwinowie, Oddział w Krakowie, Kraków.

Ragland, J. (1996) Programy rozwoju obszarów wiejskich jako część działania doradztwa rolniczego w USA (Rural development programmes as a part of agricultural extension in the USA). In: *Doradztwo rolnicze jako ogniwo systemu wiedzy rolniczej w procesach modernizacji wsi i rolnictwa oraz integracji z Unią Europejską* (Agricultural Extension as a Link of the Agricultural Knowledge System in the Process of

Modernization of Rural Areas and Agriculture and Integration with the European Union), International Conference, June 1996, pp. 12–14. Mazowiecki Osrodek Doradztwa Rolniczego w Warszawie (Mazovian Extension Service), Oddzial Poświętne (Branch in Poświętne).

Retzlaff, R. (2004) Agritourism zoning down on the farm In: *Zoning Practice*. American Planning Association, pp. 2–6. [in:] http://www.planning.org/zoningpractice/2004/pdf/mar.pdf (accessed April 2008).

Roberts, L. and Hall, D. (2001) *Rural Tourism and Recreation*. CAB International, Wallingford, UK.

Santrock, J. (1998) *Adolescence*. McGraw-Hill, Boston.

Schmidt, A. (1970/71) *Fremdenverkehr, Multiplikator und Zahlungsbilanz*. Jahrbuch für Fremdenverkehr, 'dwif e.V.' Munich.

Senauer, B., Asp, E. and Kinsley, J. (1991) *Food Trends and the Changing Consumer*. Egan Press, St Paul, Minnesota.

Shadbolt, N. and Martin, S. (2005) *Farm Management in New Zealand*. Oxford University Press, South Melbourne.

Sharpley, R. and Sharpley, J. (1997) *Rural Tourism: an Introduction*. Nelson Thomson Learning, London and Boston.

Sikora, J. (1999) *Organizacja rynku turystycznego* (Organization of the Tourism Market). Wydawnictwa Szkolne i Pedagogiczne Spółka Akcyjna, Warsaw.

Veer, M. and Tuunter, E. (2005) *Rural Tourism in Europe. An Exploration of Success and Failure Factors*. Stichting Recreatie, Expert and Innovation Centre, the Hague, the Netherlands.

Whitby, M. (1994) Promoting rural development through rural employment projects and policies: some British experience. Paper for the Rural Realities Conference, University of Aberdeen, Scotland.

Wilson, J., Thilmary, D. and Sullins, M. (2006) Agritourism: a potential economic driver in the rural West, 8 pp., Cooperative Extension, Colorado State University, Department of Agricultural and Resource Economics, Fort Collins. Available online at http://dare.agsci.colostate.edu/csuagecon/extension/pubstools.htm, accessed April 2008.

Wolfe, K. and Holland, R. (2005) Considering an agritainment enterprise in Tennessee? University of Tennessee Extension publication PB1648. Available online at http://www.utextension.utk.edu/publications/pbfiles/pb1648.pdf, accessed March 2005.

Wysocki, F. and Lira, J. (2003) *Statystyka opisowa* (Descriptive Statistics). Agricultural University of Poznań, Poznań.

Yeoman, J. (2000) The importance of rural tourism and agritourism in rural development. *Roczniki Naukowe Stowarzyszenia Ekonomistów Rolnictwa i Agrobiznesu* (Annals of the Polish Association of Agricultural and Agribusiness Economists) II (1), 41–51.

Statistical Materials, Official Journals and Documents

Agenda 2000, For a Stronger and Wider Union (1997) Bulletin of the European Union, European Commission, Brussels, Luxembourg.

Agritourism in focus, A Guide for Tennessee Farmers (2005) Centre for Profitable Agriculture. University of Tennessee – Farm Bureau Partnership. Available online at http://extension.tennessee.edu/publications/marketing/default.asp#value, accessed April 2008.

Area under permanent grassland in utilized agricultural area in the European Union (2001) European Environment Agency (EEA), Copenhagen. Available online at http://dataservice.eea.europa.eu/atlas/viewdata/viewpub.asp?id=558, accessed March 2008.

Atlas samochodowy Polski (Car Atlas of Poland) (1999) Polskie Przedsiębiorstwo Wydaw-
    nictw Kartograficznych PPWK im. Eugeniusza Romera, Warsaw.
Australian Bureau of Statistics, Survey of income and housing costs, 2000-2001. In: Rural
    Households' Livelihood and Well-Being, Statistics on Rural Development and Agri-
    culture Household Income, The Wye Group Handbook, www.fao.org/statistics/rural/,
    UNITED NATIONS, New York and Geneva, 2007, p. 392.
Buckwell Report (1997) Towards a Common Agricultural and Rural Policy for Europe
    (Report of an Expert Group). Commission of European Communities, Directorate
    General VI/A1, European Commission. Available online at http://ec.europa.eu/agri-
    culture/publi/buck_en/index.htm, accessed April 2008.
Compendium of Tourism Statistics (2003) *World Tourism Barometer* 1 (1), June, World
    Tourism Organization.
Cork Declaration (1996) *A Living Countryside.* European Conference on Rural Develop-
    ment. In: Agriculture – Rural development – The Cork Declaration, 9 November
    1996. Available online at http://www.europa.eu.int/comm/dg06/rur/cork_en.htm.
Demographia World Urban Areas: Population Projections 2007 and 2015 (2008) Avail-
    able online at http://www.demographia.com/db-worldua2015.pdf, accessed April
    2008.
Demographia. World Urban Areas (World Agglomerations), 4th Comprehensive Edition:
    Revised, August 2008. Available online at http://www.demographia.com/db-world-
    ua.pdf (accessed October 2008).
Earth Trends (2004) *Data Tables: Biodiversity and Protected Areas.* United Nations Envi-
    ronment Programme – World Conservation Monitoring Centre (UNEP – WCMC).
    Database on Protected Areas (WDPA). [In:] http://sea.enep-wcmc.org/wdpa/down-
    load/wdpa 2004/index.html (accessed March 2008). Cambridge, UK.
Economic Research Service of US Department of Agriculture. In: Income of the agricul-
    ture household sector, 2001 report, Eurostat; Rural Households' Livelihood and
    Well-Being, Statistics on Rural Development and Agriculture Household Income,
    The Wye Group Handbook, www.fao.org/statistics/rural/, United Nations, New York
    and Geneva, 2007, p. 338.
Employment in agriculture and in the other sectors: structures compared (2007) The 2007
    Agricultural Year, Eurostat. Available online at http://ed.europa.eu/agriculture/
    agrista/2007/table_en/en351.htm, accessed April 2008.
European Charter for Rural Areas (1996) Committee on Agriculture and Rural Develop-
    ment, Council of Europe.
Farm Household Income (2004) *Towards Better Informed Policies, Policy Brief.* OECD
    Observer, OECD, Paris [In:] http://www.oecd.org/dataoecd/43/54/33817664.pdf
    (accessed April 2008).
Future of Rural Society (1988) COM (88) 501 Final of 28 July 1988, EU Commission,
    Brussels. Available online at http://aei.pitt.edu/5214/, accessed April 2008.
Green Paper 'Perspectives for the Common Agricultural Policy' (1985) Communication of
    the Commission to the Council and the Parliament, Commission of the European
    Communities, COM (85) 333 final, Brussels, 15 July 1985. Available online at aei.
    pitt.edu/931/01/perspectives_for_cap_gp_COM_85_333.pdf, accessed April 2008.
Income of the agriculture household sector, 2001 report, Eurostat. In: Rural Households'
    Livelihood and Well-Being, Statistics on Rural Development and Agriculture House-
    hold Income, The Wye Group Handbook, www.fao.org/statistics/rural/, United Na-
    tions, New York and Geneva, 2007, p. 336.
International Comparison Program (2005) Tables of final results. International Bank for
    Reconstruction and Development/World Bank. Available online at siteresources.
    worldbank.org/ICPINT/Resources/ICPreportprelim.pdf, accessed April 2008.

Most Beautiful Villages . . . (2008) The savvy traveller from Chicago, Illinois. Available online at http://www.thesavvytraveller.com/insights/series/most_beautiful_villages/1home.htm, accessed April 2008.

New Perspectives for EU Rural Development (2006) Fact Sheet, European Commission, Agriculture Directorate-General, Rural Development. Available online at http://europa.eu.int/comm/agriculture/rur/index_en.htm, accessed April 2008.

New York City Zoning (2008) New York City Department of City Planning. Available online at http://www.nyc.gov/html/dcp/html/zone/zonehis.shtml, accessed April 2008.

Official Journal of the European Communities (1984) Community Policy on Tourism-Initial Guidelines C 115/02, of 30 April 1984.

Official Journal of the European Communities (1992) No. L 231, of 13 June 1992.

Opinion of the Committee of the Regions of 16 January 1997 on a Rural Development Policy (1997) European Union, Committee of the Regions, Brussels, Belgium. Available online at www.cor.eu.int/coratwork/comm2/english/cdr389-1996_fin_ac_en.html, accessed April 2008.

Opinion of the Committee of the Regions on a Policy for the Development of Rural Tourism in the Regions of the European Union (1995) Subcommission 2, 2 February 1995, Brussels Belgium. Available online at www.cor.eu.int/coratwork/comm2/english/19-1995_en.html, accessed April 2008.

Opinion of the Committee of the Regions on the Commission Green Paper on the role of the Union in the field of tourism (COM(95)97 final) (1995) Sub-commission 2, Brussels, 16 November 1995. Available online at www.cor.eu.int/coratwork/comm2/english/cdr376-1995_ac_en.html, accessed April 2008.

Report on the Results of the National Agricultural Census 1996 (Raport z wyników Powszechnego Spisu Rolnego 1996) (1998) Systematics and characteristics of farms (Systematyka i charakterystyka gospodarstw rolnych). GUS (Central Statistical Office), Warsaw.

Report on the Results of the National Agricultural Census 2002 (Raport z wyników PSR 2002) (2003). GUS (Central Statistical Office), Warsaw.

Rural Development in the European Union (2007) Statistical and Economic Information Report 2007, European Union, Directorate-General for Agriculture and Rural Development, November 2007. Available online at http://www.bookshop.europa.eu/eGetRecords, accessed April 2008.

Rural Development Policy 2007–2013 (2007) European Commission, Agriculture and Rural Development. Available online at http://ec.europa.eu/agriculture/rurdev/index_en.htm, accessed April 2008.

Rural Developments (1997) CAP 2000, Working Document, European Commission, Directorate General for Agriculture (DG VI), July 1997. Available online at http://www.europa.eu.int/comm/agriculture/publi/pac2000/rd/rd_en.pdf, accessed April 2008.

Rural Households' Livelihood and Well-Being (2007) Statistics on Rural Development and Agriculture Household Income, Wye Group Handbook, United Nations, New York and Geneva. Available online at www.fao.org/statistics/rural/, accessed April 2008.

Statistical Yearbook of Poland, 1965, Central Statistical Office, Warsaw.

Statistical Yearbooks of the Republic of Poland 1990–2007 (Roczniki statystyczne Rzeczpospolitej Polskiej 1990–2007). Central Statistical Office (Główny Urząd Statystyczny), Warsaw.

Strategy of eco-development in a rural commune (1999) Eco-development vs Agenda 21 in rural areas. Available online at http://server.eko.wroc.pl, accessed 2003.

The Austrian Farm Holidays Association. Available online at http://www.farmholidays.com/bundesverband/qualitaet.html?L=&id=1&L=3 (accessed April 2008).

The distribution of national and scenic parks in Poland in 2006, Ministry of Environment of Poland, http://www.mos.gov.pl/soe_pl/rys16b.htm (accessed April 2008).

The dominant landscape types of Europe, European Environment Agency (EEA), http://www.eea.europa.eu, Copenhagen, 2005. Available online at http://dataservice.eea.europa.eu/atlas/viewdata/viewpub.asp?id=2573 (accessed March, 2008).

The European Conference on Rural Development (1996) The Cork Declaration – a driving countryside, 9 November 1996. Availble online at: http://ec.europa.eu/agriculture/rur/cork_en.htm (accessed April 2008).

The Facts & Figures (2008): the World Tourism Organization. Available online at http://www.world-tourism.org/facts/menu.html (accessed April 2008).

Treaty of Rome (1957) The Treaty establishing the European Economic Community (EEC), Article 130A. Available online at http://europa.eu/scadplus/treaties/eec_en.htm, accessed April 2008.

Tourism Satellite Accounting Tool, World Travel and Tourism Council, London. Available online at http://www.wttc.travel/eng/Tourism_Research/Tourism_Satellite_Accounting_Tool.

Tourist Services Act of 29 August 1997 on tourism services (Official Journal No. 133, Pos. 884) published in Official Journal of 2001 No. 55, Pos. 578.

Towards Quality Rural Tourism: Integrated Quality Management (IQM) of Rural Destinations (2000) Integrated Quality Management in Rural Tourism Destinations. European Commission, Luxembourg.

Travel and Tourism: a Job Creator for Rural Economies (2000) The World Travel and Tourism Council, Travel and Tourism Millennium, London, UK.

US Census Bureau Current Population Survey 2007 (2007) Annual Social and Economic Supplement. Available online at http://www.census.gov/cps/, accessed April 2008.

USDA Rural Development (2007) Progress Report, Bringing Broadband to Rural America. Leading the Rural Renaissance. Connecting to the New Assets. USDA. Available online at http://www.rurdev.usda.gov/rd/pubs/progress/2007_RD_ProgressReport.pdf, accessed April 2008.

Wieś polska zaprasza (Polish Country Invites), Polska Federacja Turystyki Wiejskiej 'Gosopdarstwa Gościnne'. Available online at www.agritourism.pl (accessed April, 2008).

World Database on Protected Areas (WDPA) (2007), World Conservation Monitoring Centre (UNEP – WCMC) and the IUCN World Commission on Protected Areas (31 January 2007). Available online at http://sea.unep-wcmc.org/wdbpa/PA_growth_chart_2007.gif, accessed March 2008.

World POPClock Projection (2008) International Programs Center, U.S. Census Bureau. Available online at http://www.census.gov/ipc/www/popclockworld.html (accessed April 2008).

World Resources (2005) *The Wealth of the Poor: Managing Ecosystems to Fight Poverty.* UN Development Programme, UN Environment Programme, World Bank, World Resources Institute, Washington, DC.

World Tourism Barometer I (1), June 2003, and *World Tourism Barometer* 5 (2), June 2007. World Tourism Organization.

World Tourism Barometer 6, (1), January 2008, World Tourism Organization. Available online at http://www.unwto.org/facts/eng/pdf/barometer/UNWTO_Barom08_ 1_en.pdf (accessed April 2008).

World Urbanization Prospects (2005) The 2005 Revision. United Nations Department of Economic and Social Affairs/Population Division, New York, NY.

Other Sources

Agritourism (2006) The beyond organic show. Available online at http://www.beyondor-ganic.com/template/nst.php?id=081705&idy=2005&sn=sn2, accessed April 2008.

Agritourism in Europe (2004) Mintel International Group, London, UK.

Good to Grow Farmers Enhance Their Prosperity with Agritourism (1997) *Arizona Business Gazette*, 27 March 1997.

Improving Competitiveness (2008) Quality in Tourism: A Conceptual Framework, World Tourism Organization, Committed to Tourism, Travel and the Millennium Development Goals. Available online at http://www.unwto.org/quality/std/en/std_01.php?op=1&subop=1, accessed April 2008.

Integrated Quality Management in Rural Tourism Destinations (2000) Available online at http://www2.ceredigion.gov.uk/english/visiting/tourism/certwg/ECIQMRuralDigest.htm#olnod, accessed April 2008.

Mapa Agroturystyczna woj. zachodniopomorskiego (An Agritourist Map of West Pomerania Province) (2003) Koszalińskie Stowarzyszenie Agroturystyczne KOSA (Koszalin Agritourist Asscociation), Koszalin.

New Zealand Herald of 26 November 2003. A11.

Quality Assessment Standards (2005) Rural Accommodation. European Federation for Farm and Village Tourism 'EuroGites'. Available online at http://www.eurogites.org/documents, accessed April 2008.

US Census Bureau Current Population Survey (2007) Annual Social and Economic Supplement.

VisitBritain (2008) Who's Who in Quality. Available online at http://www.qualityintourism.com/asp/main.asp, accessed April 2008.

Internet Sources

Agritour, New Zealand's Specialist Agricultural Tour Company: http://www.agritour.co.nz/index.html.

AgriTours Canada Inc.: www.agritourscanada.com.

Agri-Travel International (a branch of Calder and Lawson Travel): http://www.agritech.org.nz/agritravel.shtml.

Agrotravel: http://www.agrotravel.pl.

Amazing Maize Maze online: www.americanmaze.com.

@home New Zealand, the official site of the New Zealand Association of Farm and Home Hosts: http://www.athome.org.nz.

Austrian Farm Holidays Association: http://www.farmholidays.com.

Barossa South Australia: http://www.barossa-region.org.

Delta Society: http://www.deltasociety.org/index.htm.

Dolphin-Assisted Therapy (DAT): http://www.dolphinassistedtherapy.com.

Eden Project: www.edenproject.com.

Enjoy England, 'VisitBritain': http://www.enjoyengland.com/about.

EuroGites, the European Federation for Farm and Village Tourism: http://www.eurogites.org.

FAOSTAT: http://faostat.fao.org.

Farm to Farm New Zealand Tours: http://www.farmtofarm.co.nz.

Ferien auf dem Bauernhof: http://www.bauernhof-ferien.ch.

Food and Agriculture Organization of the United Nations: http://www.fao.org.

Gîtes de France: http://www.gites-de-france.com/gites/uk/rural_gites.

Google Earth: http://earth.google.com/intl/pl.

Hospitality Plus, the New Zealand Home and Farmstay Co. Ltd: http://www.ruralhols.
co.nz.

Kings Ranch: http://www.kings-ranch.com.

Manitoba Country Vacation Association: http://www.countryvacations.mb.ca.

Ministry of the Environment: http://www.mos.gov.pl/index_main.shtml.

National Parks Worldwide: http://www.staff.amu.edu.pl/~zbzw/ph/pnp/swiat.htm.

New Zealand Farm Holidays Ltd: http://www.nzfarmholidays.co.nz/images/title.jpg.

Pets and People's Feline AAT Program: http://www.petsandpeople.org/cat-aat.htm.

Polish Federation of Rural Tourism 'Hospitable Farms': http://www.agroturystyka.pl/in-
dex; www.agritourism.pl.

Polish Institute of Tourism: www.intur.com.pl.

Polish National Parks: http://www.staff.amu.edu.pl/~zbzw/ph/pnp/pnp.htm.

Rural Tourism Holding (RTH) http://www.ruraltourism.co.nz/default.htm.

Savvy Traveller: www.thesavvytraveller.com.

South Australian Commission: http://www.barossa.com/site/page.cfm.

Tourisme Côte de Beaupré: www.cotedebeaupre.com.

Urlaub am Bauernhof: http://www.urlaubambauernhof.com.

US Census Bureau: http://www.census.gov.

Wikipedia: http://en.wikipedia.org.

World Travel and Tourism Council: http://www.wttc.travel.

World Travel and Tourism Council, Tourism Satellite Accounting Tool: http://www.wttc.
travel/eng/Tourism_Research/Tourism_Satellite_Accounting_Tool.

Zanzibar.net – Your Passport to Paradise: http://www.zanzibar.net.

Index